Grassroot Innovations in Fisheries and Aquaculture

THE AUTHORS

Dr. Arpita Sharma is Principal Scientist and Head of Fisheries Economics, Extension and Statistics Division/Social Sciences Division at Indian Council of Agricultural Research-Central Institute of Fisheries Education, a Deemed University, located in Mumbai, India.

Dr. Shivta Kureel completed her Doctorate in Fish Business Management from ICAR-Central Institute of Fisheries Education, Mumbai. Her research area was grassroot innovations in the fisheries sector. For here research she worked passionately to collect various grassroot innvoations from all over India. She is also an experienced educator (primary) and ex fellow from 'Teach For India'.

Grassroot Innovations in Fisheries and Aquaculture

Authors: Arpita Sharma | Shivta Kureel

Editors: Rupam Sharma | S.P. Shukla

Artist: S.K. Sharma

ASTRAL
Publishing Authors Across the Globe

2025

Daya Publishing House®
A Division of
Astral International Pvt. Ltd.
New Delhi – 110 002

ISBN: 978-93-5919-916-0

Published by : **Daya Publishing House®**
A Division of
Astral International Pvt. Ltd.
– ISO 9001:2015 Certified Company –
4736/23, Ansari Road, Darya Ganj
New Delhi-110 002
Ph. 011-43549197, 23278134
E-mail: info@astralint.com
Website: www.astralint.com

Printed at : **Replika Press Pvt. Ltd.**

Cover page pics are AI generated

ACKNOWLEDGEMENT

Importance of innovation is illustrated by the fact that 2010-2020 was declared' decade of Innovation' in India. In fisheries sector, countless new ideas and innovations are applied regularly but are rarely documented.

It is imperative to understand and draw lessons that can explain the dynamics of the innovation syste ICAR-CIFE has been in forefront to document the Indigenous Technical Knowledge (ITKs) in Fisheries sector by organizing writeshops in a nationwide project. During the conduct of these writeshops, it was realized that the grassroot innovations being practiced in the present times were not qualifying in the definition of ITKs. Thus, it was necessary to document grassroot innovations. In July, 2013, under institutionally funded research project "Documenting Grassroot Innovations in Fisheries Sector and Facilitate Patenting of Innovations by CIFE Scientists" first Student Research Convention on Innovations in Fisheries was organized at CIFE by the project team and PGSSU. In this convention, 195 participants including students from fisheries science, researchers and scientists from over India bestowed their knowledge and contributed in documentation of grassroot innovations. Selected innovations have been documented in this book with due credits to the farmer/fisher practicing the innovation as well as to the persons who have documented it.

Arpita Sharma
Rupam Sharma
S.P. Shukla
Shivta Kureel

PREFACE

ICAR-CIFE believes that today's youth and students have an outstanding ability to grasp, understand and accept any change that is taking place in our system and many times, they want to share it with the world. Realizing the importance of innovation, 'First Students' Research Convention on Innovations in Fisheries' was organized by ICAR-Central Institute of Fisheries Education (CIFE), Mumbai and the Post Graduate School Students Union (PGSSU) of CIFE, Mumbai on the occasion of National Fish Farmers' Day on 10th July, 2013 when Dr. W.S. Lakra was the Director ICAR-CIFE, Mumbai.

The objectives of the convention were to document innovations in fisheries sector and to provide them a platform to assemble, discuss innovations collected from field and share it at a national level convention. There were 195 young professionals from 17 states belonging to 20 colleges as well as from Sudan and Eritrea, Africa who assembled and presented the innovations which they have seen in fields, in the form of creative posters. Students interacted with farmers, scientists, researchers and academicians. More than 150 innovations were documented covering the field of aquaculture, aqua tourism, cage culture, capture fisheries, aquatic health, aquarium management, pond preparation, fish processing, fish transportation and marketing, integrated farming, fish harvesting and related fields. Based on the contributions, a resource book on "Grassroot Innovations in Fisheries and Aquaculture" has been prepared which has a collection of more than 60 innovations. After this convention, more information was collected by interviews and visits.

Citation: Arpita Sharma and Shivta Kureel (2025) Grassroot Innovations in Fisheries and Aquaculture, Rupam Sharma and S.P. Shukla (eds), Published by Daya Publishing House®, A Division of Astral International Pvt. Ltd., Ansari Road, Darya Ganj New Delhi-110 002.

Arpita Sharma
Rupam Sharma
S.P. Shukla
Shivta Kureel

CONTENTS

1

Aeration Using Bamboo Baskets

Aeration of ponds through mechanical devices escalates the cost of production in aquaculture. To avoid such extra costs, farmers have developed a low cost innovative device for maintaining adequate oxygen level in fishponds using bamboo as raw material. A tripod stand of bamboo is fixed in water, keeping two third portion of the stand above water surface. Three bamboo baskets are fixed to the stand at an interval of one meter each. The lower most basket is approximately 1 meter above the surface of water. A motor is installed on the dyke with inlet pipe called suction end and an outlet pipe called delivery end. A flexible pipe is connected to delivery end, which leads to topmost basket. Water from pond is sucked in and release out through delivery pipe in the topmost basket. Water trickles down from the top basket to the lower two baskets and finally on to the water surface. This provides sufficient aeration of water falling from the vertically placed baskets. This process helps in maintaining the adequate level of dissolved oxygen and water circulation in ponds.

Practiced: Farmers of Odisha

Documented by: Nibedita Kalita, M.F.Sc. Student, ICAR-CIFE, Mumbai

Email: nivedita.p13@gmail.com

2

Automatic Under-Gravel Filtration System for Water Purification

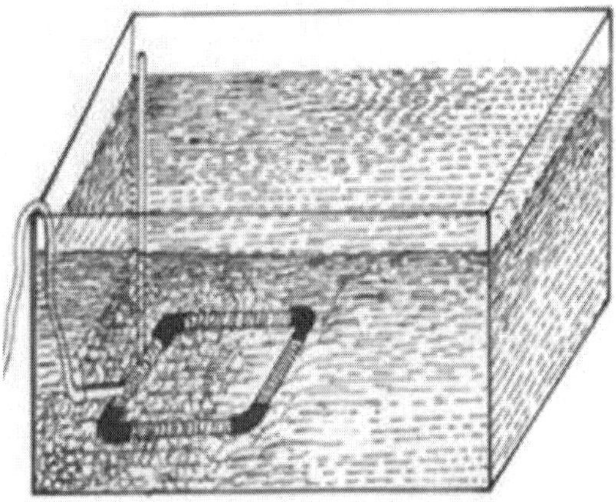

Automatic under-gravel filtration system for aquaria, PVC tanks and cemented pools is a low cost system for effective cleaning of water. It is a type of indigenous water filter made of locally available materials which are cheap, and easily available even in villages and remote areas. Construction of this filtration device is very simple, where a square/ rectangle frame of PVC pipes is made with the help of four PVC elbow and a "T" shaped PVC pipe which is joined to one of the PVC sides. An aeration tube is connected to "T" shaped PVC pipe for creating airlift mechanism. Slits are made on each side of rectangular frame, which allows only water to sweep into PVC pipes. This structure is placed inside aquarium or cemented tub, which is filled with gravels of appropriate size. This entire structure act as filtration system and it is placed inside a bigger aquarium/tank or cemented pool for water purification. The principle behind this system is that as air moves up through the PVC pipe, it acts as air lift pump which purifies water leaving dirt and debris accumulated on the gravel bed. The slits on PVC pipes allow only the water to pass through, whereas the gravel act as cleaning system allowing only water to trickle down. Debris/dirt is later removed using suction tube or siphoning bottle. It is an efficient way of cleaning water in aquarium, tanks and cemented pools leaving water crystal clear

Practiced by: Dr. S.G.S. Zaidi, Technical Officer, ICAR-CIFE, Mumbai.

Email: sgszaidi@hotmail.com

Documented by: Shivta Kureel, Ph.D. Scholar, CIFE, Mumbai

Email: kureelshivta@yahoo.com

3

Enclosure Made of Bamboo Frame for Better Fish Feed Utilization

While feeding fish, spreading of feed is a common problem observed in fish ponds, because of which a considerable quantity of feed is wasted resulting in higher production cost. To avoid this waste of feed, fish farmers have developed a feed distribution device which is a rectangular frame made of bamboo. This hollow structure/enclosure is fixed in the pond generally at the centre. The feed is broadcasted within this enclosed frame at particular timings to ensure that fish develops a habit of taking feed from the enclosed structure at the set time daily. Feed spreads within this enclosure ensuring most of the feed is utilized by fish. Fishes become habitual of taking feed from this specific area and aggregate there at a fixed time, this allows proper utilization of feed and thereby reducing the cost of production.

Practiced by: Farmers of Bihar and Shri.Nakul Sadafule, Kalamboli, Navi Mumbai

Documented by: Shweta Kumari, M.F.Sc. Student, ICAR-CIFE, Mumbai

Email: shwetaniki88@gmail.com and Rashmi S. Ambulkar, Ph.D. Scholar, CIFE, Mumbai

4

Collection of Prawn Seed From the Dam Slope Site

In nature, gravid female prawn move to estuary for purpose of breeding and laying eggs after which, spent female migrates back to freshwater. Also, the post larvae or prawn at juvenile stage migrate towards the freshwater. Fishers from Dhowlashwaram Barrage in Andhra Pradesh collect these juvenile from the dam slopes site from river Godavari during migration from estuaries to freshwater. A net is fixed down at these sites from where juveniles are trapped and collected. Prawn by habit climb up the dam slopes and are trapped in nets. However, disadvantage of this practice is that there is a mixture of juvenile catch, which comprises of various species of prawns.

Practiced by: Fishers of Andhra Pradesh

Documented by: Bharti Huriem, M.F.Sc. Student, ICAR-CIFE, Mumbai and Dr. Asha Landge, Senior Scientist, CIFE, Mumbai

Email: ashalandge@yahoo.com

5

Cemented Jar Hatchery for Breeding of *Pangasius Sutchi*

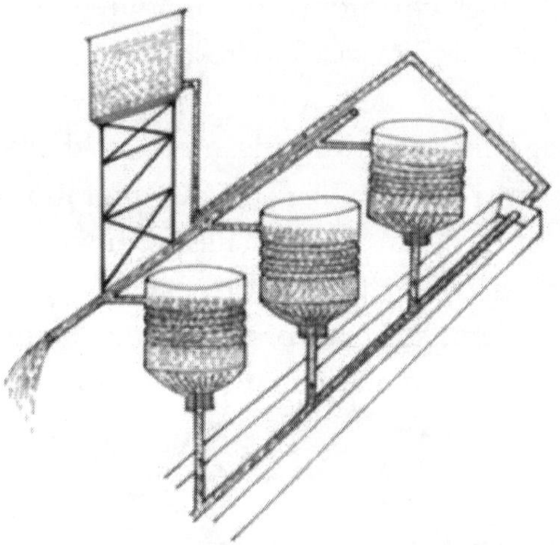

A cemented hatchery for *P. sutchi* is constructed which works on the principle of glass jar hatchery and is being used by farmers of Raipur, Chattisgarh. Each unit of hatchery called as "cemented jar" is a cylindrical shaped structure having a conical base attached at the bottom. This hatchery has 16 cemented jars placed in a platform and three breeding tanks. Apart from construction cost, another major advantage is the durability of this structure.

In this hatchery the inlet of an individual cemented jar is at the bottom. Due to upward pressure of water current released in the jar, creates. vertical movement is For complete drainage of each unit, inlet pipe is removed and water is completely drained through it. The outlet is placed about 3 inch below the top rim from where excess water is removed. In this type of structure *P. sutchi* fish can be bred and hatched.

Practiced by: Fish farmers of Raipur, Chattisgarh

Documented by: Shivta Kureel, Ph.D Scholar, CIFE, Mumbai and Dr C.S. Chaturvedi, Senior Scientist, ICAR-CIFE, Mumbai, **Email:** chaturvedi@cife.edu.in

6

Water Cleaning Device Made of Used Plastic Bottles

Siphoning has been carried out since ages for cleaning aquarium. But while spiphoning, dirt and debris, small fish also enters in to the suction pipe due to high water pressure. To avoid such problem, an innovative device using used plastic bottle has been developed. An old plastic bottle of soft drinks or packaged water, preferably of 2 litres capacity, is cut from the bottom end forming a wide circular mouth. The corked end is connected to locally available plastic pipe which fits perfectly into the screwed end of bottle forming the head of the structure. The working principle behind this is that the high pressure caused due to suction tube is distributed uniformly over the wide mouth which allows only dirt and debris to pass through the mouth.

Gravel, because of its weight, remains near the mouth and settles down to its bed without passing through the head of the bottle. Hence, only water, dirt and debris are allowed to be siphoned out of aquarium, tank or cemented pool, keeping water relatively cleaner.

Practiced by: Dr. S.G.S. Zaidi, Technical Officer, ICAR-CIFE, Mumbai

Email: sgszaidi@hotmail.com

Documented by: Shivta Kureel, Ph.D. Scholar, ICAR-CIFE, Mumbai

Email: kureelshivta@yahoo.com

7

Live Fish Transport in Auto Rickshaws and Cycle Rickshaws Modified by Plastic Material

A rectangular deep container made of thick synthetic material (*tripal*) is used inside an auto rickshaw for transporting fish. *Tripal* is tied behind the driver seat and the passenger seating area. In case of cycle rickshaw, it is converted into a simple transportation means either by replacing passenger seat by a rectangular wooden box which is covered by two locally available polythene sheets or utilizing / converting leg space of rickshaw into a temporary water proof container using *tripal,* tied on end poles. It is generally used for transportation within a short distance. Water is filled inside the container and live fish are kept for transportation.

Water is changed in between if required. Cycle rickshaw is generally used by small scale farmers. The advantage of this innovation is that the fish can be taken back from market if not sold at appropriate price. It is an easy, quick and low cost transportation method. Live fishes fetch higher price (approximately 10% more) than dead fish. *Tripal* is easy to be installed and detached as it is a temporary structure. It is an additional source of income for rickshaw drivers as well.

Practiced by: Fishers of Bombay Fish Market, Punjab and Haryana

Documented by: Shri. Brijesh Kumar, Ph.D. Scholar, ICAR-CIFE, Mumbai

Email: brijesh.fec011@cife.edu.in

8

Multistorey Magur Hatchery

Due to irregular and erratic electricity supply in some places, insufficient and deferred water circulation can exist, resulting in cannibalism and mortality of fish. To reduce this, a magur hatchery has been built over a cemented house. On the first floor a magur hatchery is built over a surface of 60 feet X 40 feet area. On the second floor a big overhead tank of equivalent area is constructed which provides water for 3 to 4 days. It is designed for automatic circulation of water without use of motor. Overhead tank is built by extending already existing roof wall by 1 meter height. The floor of roof and the surrounding walls are thickened and plastered properly to prevent cracking or bursting of walls due to the water pressure. Water circulation is also used for maintaining water quality and providing aeration.

Practiced by: Shri.Baghel (Baghel Fisheries Farm), Raipur, Chhattisgarh

Documented by: Shri.Brijesh Kumar, Ph.D. Scholar, ICAR-CIFE, Mumbai

Email: brijesh.fec011@cife.edu.in

9

Preparation of Fish Feed Using Vegetable Peels and Rotten Rice

Rotten rice grains in most of the houses are usually thrown away. Farmers of Nadia district, West Bengal use these rotten rice grains along with vegetable peels as fish feed. For this, the rotten rice grains are kept in sacs along with vegetable peels. These sacs are submerged in water for 7-12 days so as to decompose the rice and vegetable peels. These decomposed matters are used for fish feeding.

Practiced by: Shri.Sajrul Haque, Fisher, North-24 Parganas, West Bengal

Documented by: Shri.Amitava Ghosh, M.F.Sc. Student, ICAR-CIFE, Mumbai

Email: amitava.skg@gmail.com

10

Trapping Solar Energy Through Plastic Bags to Raise the Temperature in Winter Season

During winter season, growth of fish reduces and extreme low temperature cause fish mortality. Covering the entire pond surface by water filled plastic bags provides a low-cost solution to this problem. Locally available polythene bags or even good quality plastic bags (more durable) can be used for this purpose. Water from pond itself or preferably saline water (sea water or water mixed with locally available salt) is filled in these plastic bags up to one third of the capacity of the plastic bag. Remaining two third of the plastic bag is inflated manually and tied using locally available rope/ thread. These inflated bags are placed over the water surface which float on the surface. The salt water traps energy from sunshine and raise its temperature simultaneously. This energy once trapped is unable to escape into environment and hence slowly diffuse into pond/tank water. These plastic bags are placed tightly together so that there is no open space in between them forming a layer of blanket over the pond/tank surface. The temperature can be increased up to 5-100C by this method. It is a cheap and effective technology to maintain the water temperature during winter and also ensures survivability and proper growth of the fish.

Practiced by: Dr. S.G.S. Zaidi, Technical Officer, ICAR-CIFE, Mumbai

Email: sgszaidi@hotmail.com

Documented by: Shivta Kureel, Ph.D. Scholar, ICAR-CIFE, Mumbai

Email: kureelshivta@yahoo.com

11

Use of Alum to Prevent Turbidity

Turbidity of water is a common problem in the farmed ponds. High turbidity decreases primary productivity of the pond which is not desired for fish growth. To control such turbidity, alum is used. Alum is mixed with water and spread into water. However, in larger water bodies spreading of the alum becomes difficult. To overcome such problem, farmers of Bihar have developed an innovative method to spread alum. Alum is tied to both the sides of a boat and as the boat moves in the pond, alum get dispersed resulting settle down of the debris and thus reducing the turbidity of the pond.

Practiced by: Farmers of Bihar

Documented by: Shweta Kumari, M.F.Sc. Student, ICAR-CIFE, Mumbai

Email: shwetaniki88@gmail.com

12

Use of Broken PVC/Cement Pipes as Hideouts for Rearing *Clarias Batrachus* in Ponds

The Asian catfish, *Clarias batrachus* (magur) is one of the most important aquaculture species. Magur needs hide outs in the culture pond as this species has a habit of taking shelter. Hence, to simulate the natural habitat for the fish, farmers of Andhra Pradesh use either broken PVC pipes or cement pipes in the ponds. These pipes are easily available in market. Unused household pipes can also be used for this purpose. They act as hideouts or shelter. In natural habitat, magur chooses rocks, stone or boulders for shelter or hideouts. However, in controlled system PVC pipes or cemented pipes are preferred.

Practiced by: Farmers of Andra Pradesh, India

Documented by: Bharti Hureim, M.F.Sc. Student, ICAR-CIFE, Mumbai and Dr. Asha Landge, Senior Scientist, CIFE, Mumbai,

Email: ashalandge@yahoo.com

13

Use of Compact Disc, Polythene Strips and Wrapping Paper as Bird Scaring Device in Fish Farm

In order to prevent the loss of fish due to birds (cormorants), an efficient and effective way has been innovated by farmers. Farmers tie nylon threads or other materials across the fish ponds using bamboo or wood stands and place CD at regular distance on them. When sunrays are reflected by these CDs, birds lose the sight of fish resulting in avoidance of loss due to predatory birds.

Another way to scare the brids is to tie old polythene bags or strips instead of CDs. These polythene strips are made out of old poly-bags available in market. The sound made by these strips when air blows, scare away birds, hence saving loss or damage caused by them. Wrapping paper is also used to scare the birds away by both noise and reflection of sun light. In addition, it is cost effective and has higher efficiency as it works in all seasons.

Practiced by: Fish farmers of Gujarat

Documented by: Shivta Kureel, Ph.D. Scholar, ICAR-CIFE, Mumbai and Rashmi Ambulkar, M.F.Sc. Student, ICAR-CIFE, Mumbai.

14

Use of Pipes for Maintaining Water Level in Hatchery

Maintaining a desired water level in hatchery is a common problem faced by hatchery owners. Drilling a permanent hole in the tank would not allow the tanks to hold large quantity of water and would restrict use of the tank. Using laws of physics, Mithila Matsyahatchery, Madhubani, Bihar has developed an innovative technology for maintaining water levels at various depth as per requirement of the fish species and stages of development of fish. This hatchery uses pipes for maintaining water level at desired height. The pipes are connected from the side at the bottom of the tank which allows easy drainage and complete emptying of the cemented tank. Pipes are adjusted in height according to required water level because of which extra water automatically drains out. Pipes are flexible and they can rotate upto 180 degrees, chains are used to hold pipes at required angle. After excess water is drained out, desired water level is maintained in the pond.

Practiced by: Farmers of Bihar

Documented by: Shweta Kumari, M.F.Sc. Student, CIFE, Mumbai

Email: shwetaniki88@gmail.com

15

Boro Paddy - Fish Integration in Low Lying Areas of Nagaon, Assam

The basic principle of aquaculture is to achieve higher production per unit area. For this, farmers of Nagaon district of Assam converted their boro paddy plots for extended periods for small size fish culture which is proved to be very successful. The transformed paddy plots range from 0.50 to 5.0 ha in size, with elevated dykes with a pond or large canals inside, which may retain water year round.

Normally recommended boro paddy varieties like Kanaklata, Brishnuprashsd, Joymoti, etc. are transplanted during December – February. The entire paddy plot get submerged by late April, when fry (50 – 100 mm size) of Labeo rohita, *L. gonius, L. calbasu, Catla catla, Cirrhinus mrigala, C. reba,*

Hypophthalmychthys molitrix are stocked without following species ratio into the deeper water areas. The initial stocking density happens to be 1.5 to 2.25 lakh per hectare or 9.0 quintal per hectare. Stocked fishes are fed ad libitum with fermented mustard oil caked mixed with raw cow dung and agrimin. Lime @ 18 kg every fortnight per hectare and inorganic fertilizers such as 112 kg Urea, 150 kg Super phosphate and 150 kg DAP per hectare are applied every month. Boro paddy is harvested by the month of May and the plot is ploughed to facilitate rotting of paddy straws that produces swarm of zooplankton in rising water as fish food.

After initial stocking, some of the fishes attain an average weight of 70 to 100 grams within about 90 days and are harvested. Afterwards, fishes are harvested at every two months interval with the final harvest in December – January as water dries up in the plot. Fish weighing 70 – 100 gram size have good local market value and fish growers dispose them at farm gate.

On an average 4 – 4.8 tons per hectare fish in 9 months and 4 – 5 tons per hectare boro paddy are produced annually provided the areas are inundated by rain water. In order to lure the fishes to stay backwater fishers broadcast fish feeds until water recedes. Multiple harvesting of fishes, yield comparatively higher production from such integrated culture system.

Practiced by: Md. Habibur Rahman, Godaimari village, P.O.: Juria, District, Nagaon, Assam.

Documented by: Dhiren Ch. Nath and Parag Mani Deka, M.F.Sc. Student, College of Fisheries, Assam Agricultural University, Raha, Nagaon, Assam

E mail: nathdhiren@gmail.com

16

Use of Almond Leaf (*Terminalia catappa*) in Aquarium to Lower Water pH

Terminalia catappa is a large tropical tree in the tree family, *Combretaceae* that is native to the tropical of Asia, Africa, and Australia. Indian almond leaves have been used for both therapeutic and medicinal purposes in aquarium. These leaves are said to simulate the natural environment for fish such as Bettas, Killies and Tetras. In fish keeping tanks, the leaves are also used to lower the pH and heavy metals in the water. It has been utilized in this way for Betta and Neon tetra breeding in India. Use of one almond leaf for every 15 litres of water stimulate breeding and hardening of the scales. Just like driftwood and peat, the leaves contain flavonoids (like kamferol or quercetin), tannins (such as punicalin, punicalagin or tercatin), saponines and phytosterols. Tanins play a major role to increase the acidity of the water. The tannins affect the pH-value (depending on the bufferring capacity of water) and the water turns dark. It requires 2-4 days to decompose the dry leaf in the water and it starts to secrete humic acid and different types of tannins.

Tannins also offer some apparently universal benefits like antibacterial properties and are a component in many herbs. In this respect, any source of tannic acids may benefit the aquatic environment by reducing the pH and hardness of the water. It also serves as a natural breeding habitat for fishes.

Practiced by: Shri.Nitesh Kadtan, M.F.Sc. Student, ICAR-CIFE, Mumbai

Email: ans32006@yahoo.com

Documented by: Shri.Nitesh Kadtan, M.F.Sc. Student, ICAR-CIFE, Mumbai

17

Series of Ponds Constructed on Undulating Area

In comparison to agriculture and animal husbandry, aquaculture is a new enterprise in Meghalaya.

The climate of Meghalaya varies from temperate in mid altitude (1965m msl) to sub tropical in plain zone (100m msl). Taking advantage of undulated topography, fishers of Meghalaya, particularly in hilly terrain of Khasi hills, have shown an innovative approach in rearing fishes in impoundments.

For fish culture, series of ponds are constructed on undulating area. To fill water in pond, stream water is diverted towards the first pond through the inlet.

Outlet of the first pond acts as inlet of the second pond and so on. Water flows continuously because of gravity. This allows continuous water exchange within the system. This type of innovation is seen at Laitlyngot area of Pynursla block, East Khasi hills district.

Practiced by: Farmers of Mawhati block: Umsning, district: Ri-Bhoi (Khasi hills), Meghalaya.

Documented by: Shri.Rupak Nath, Assistant Professor, Department of Fishery Science, St. Anthony's College, Shillong, Meghalaya

18

Under Water Wooden Cages for Storing Fish Catch in Wular Lake, Kashmir

Wular lake, the largest freshwater lake of Asia located in Baramullah and Bandipore districts of Kashmir valley, acts as a siltation basin for river Jehlum. The fishers living around catchment area harvest large quantity of fish from the lake particularly during summer season. The fishers sell their catch to the local middlepersons. There is intense competition among them to purchase the catch from fishers immediately after catching for which trading is done in their boats. In order to store the bulk catch, they have innovated an efficient and cost effective idea.

They construct wooden cages made of deodar tree wood which is water resistant in nature and has a longer shelf life. The cages are fixed to long bamboo poles, which are secured to the bottom of the lake. These cages are 2m deep and covered with a lid, which can be locked. The fish are stored it in these cages without incurring extra overhead charges on cold storage with the added advantage of keeping fishes in live condition in their natural environment which encourages growth and wellbeing. Each cage has a name plate of the respective fisher to whom it belongs. This innovative wooden cage reduces the cost of storage and also keep fish fresh for larger period.

Practiced by: Middlemen involved in the marketing of fish at Wular lake, Kashmir, Laherwal pora Bandipora district.

Documented by: Neha W.Qureshi, M.F.Sc. Student, ICAR-CIFE, Mumbai and Dr. M. Krishnan, Principal Scientist, CIFE, Mumbai

Email: nihaqureshi01@gmail.com

19

Use of Coral Sand to Enhance Coloration of Fish in Aquarium

The coral sand, which as the name suggests, is made up of finely ground coral, along with tiny pieces of things like seashells, calcareous algae, and sea urchin spines. Calcium carbonate, being the main component of coral sand, is a very effective buffering agent and increases pH and hardness of water in aquarium. For this reason, coral sand cannot be used in any tank with fish which cannot tolerate hard, alkaline water such as angelfish, tetras, barbs, catfish, gouramis, and loaches. On the other hand, those fish that do appreciate such conditions, such as Rift Valley cichlids, many livebearers, and brackish water fish, positively thrive in tanks with coral sand. Coral sand is relatively coarse, and so to a certain extent can be combined with other types of substrate to create an under gravel filter. A thin layer or coral sand (1-2 cm) on top of a plastic gravel will not impede the flow of water. Crushed shells or a coarse calcareous substrate such as calcium plus can be placed underneath the gravel. All these will work together and provide an excellent biological filter. It also helps in maintaining high pH and alkalinity. However, they often have the undesirable side effect of causing fading of colour fish.

Practiced by: Technocrat Aquarium, Business firm, Ulhasnagar (Thane)

Email: harshjoshicofsn@gmail.com

Documented by: Shri.Harshavardhan D. Joshi, M.F.Sc Student, ICAR-CIFE, Mumbai

Email: harshjoshicofsn@gmail.com

20

Use of Organic Slurry for Controlling Luminous Bacteria Related Problems in Shrimp Ponds

Luminous bacteria are gram- negative bacilli and facultative anaerobes which grow in semi- intensive penaeid shrimp and are reared in seawater as well as brackishwater fed shrimp culture syste Five species of luminous bacteria, viz., *Photobacterium leiognathi, Vibrio fischeri, V. harveyi, V. orientalis, and V. splendidiis* biotype 1 are encountered in seawater fed ponds. *Vibrio harveyi* is the dominant species, comprising about 80.00-92.30% of the total luminous population and claimed to be the causative agent associated with shrimp mortality. The luminous bacteria are part of the autochthonous flora of the eutrophic shrimp farm environment like that in marine environment which amplify with period of culture.

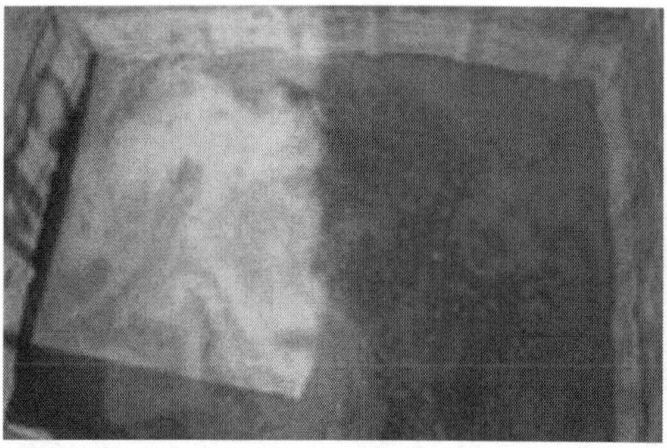

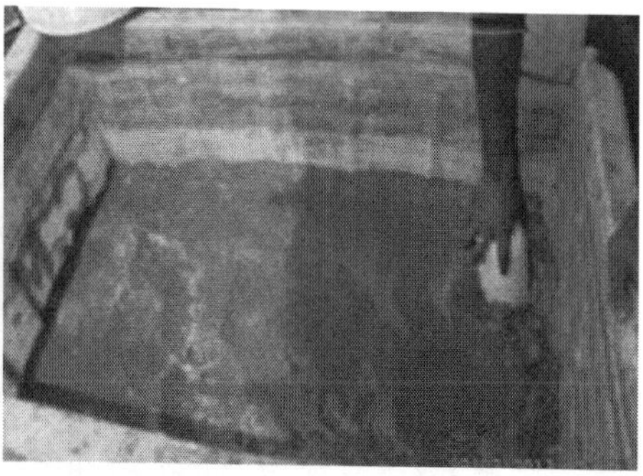

These bacteria cause bioluminescence, which is one of the forms of a chemiluminescent reaction. Luminous bacteria inhabit the gastrointestinal tract. Luminous vibriosis caused by *V. harveyi*, is a serious disease problem in shrimp aquaculture. Vibrio grow attached to algae, and may reach high population densities after being ingested with the algae and then excreted with lysed algae in faecal pellets by zooplankton. They are gut bacteria in fish and shrimps as well as zooplankton.

Under normal conditions, luminous bacteria remains in the planktonic association in water and gut through a symbiotic association in shrimp. Due to constant inputs of these bacteria expelled on faecal particles into the shrimp ponds, the association of these organisms may shift from planktonic to saprophytic or parasitic depending on the prevailing conditions. In grow-out ponds, luminousdisease frequently causes mortality with 2-3 month old stock of *Penaeus monodon* and a bacteroiphage *V. harveyi* has been implicated in mortality of such shrimp exhibiting Tea - Browngill Syndrome (TBGS). These bacteria can persist in hatchery by forming biofilms on various surfaces such as the tank surface, pipes etc.

Chlorine is widely used in hatcheries and ponds, but its use stimulates the development of multiple antibiotic resistance genes in bacteria. The vibrios surviving after chlorine treatment are not only resistant to antibiotics, but are also pathogenic. Thus the problems have been exacerbated by the use of antimicrobial compounds.

The Vibrio species are resistant to every antibiotic used, including chloramphenicol, furazolidone, oxytetracycline and streptomycin. So farmers are now using beneficial bacteria as probiotics and bacteriophage with the replacement of antibiotics. But those are expensive and less available. Probiotics such as the Gram positive Bacillus offer an alternative to antibiotic therapy for sustainable aquaculture.

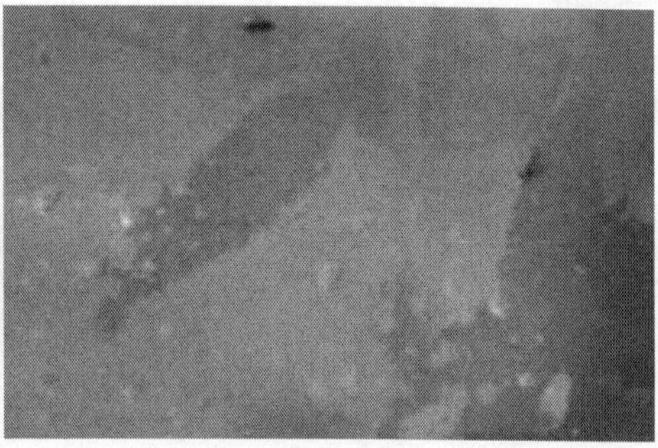

To combat such problem, marginal farmer of Purba Medinipur, West Bengal have developed a typical composition of a juice (slurry) to control Luminescent bacteria in farm ponds. The slurry is being produced by mixing 2kg. each of rice husk/wheat flour, molasses, ground nut oil cake and 200-250 gm baker's yeast in 20 l fresh water.

The mixture is kept for 3 days with occasional stirring for fermentation. When the colour turns yellow it is sieved with a small mesh size cotton cloth. The filtrate is applied directly to the pond water with an area of 1000 m2. This application is repeated after one week and the frequency of application depends upon the severity of infestation. This slurry is found to be helpful in completely eradicating gram-negative luminous bacilli, grown in semi-intensive penaeid shrimp farm, cultured in both seawater and brackishwater.

Practiced by : Fish Farmer, Medinipur, West Bengal

Documented by : Mr. Amit Mandal, M. F. Sc. Student, Dept. of Aquaculture, WBUAFS, West Bengal

21

Use of Rice Bran and Paddy Grass Rope (Hueb) for Fish Preservation and Drying

In Wular lake the majority of catch comprises of schizothoracic species. The small sized fish are used for drying. After dressing and cleaning the fish, the fisherwomen apply a layer of rice bran powder over them and the fish are hanged using Hueb (locally made ropes).

These ropes are specially designed and made of locally available dry paddy grass. The grass is plaited together and knots are made at equal distance in which the fish are placed leaving a gap of 5cm in between the ropes. These ropes with entangled fish are hung on bamboo sticks or on wooden frames for sun drying. The frames are kept at a height so that dogs and cats cannot reach the fish.

The dried fishes are sold in local markets at different rates depending upon the size. The shelf life of dried fish is 2 months. Such fishes are sold at the rate of Rs. 60- 80/kg. In order to protect the fish from any kind of infestation, the ropes are covered with polythene sheets.

Practiced by: Fisherwomen of Wular lake, Laherwal pora Bandipora district, Kashmir

Documented by: Neha W.Qureshi, M.F.Sc. Student, ICAR-CIFE, Mumbai and Dr. M. Krishnan, Principal Scientist, CIFE, Mumbai, Shri. Sheikh Imtaiz Qayoom, Sheikh Shafat Hussain, Syed Aalia Mushtaq, Iram Farooq and Sauliheen Qadri, Post Graduate Students, College of Fisheries, Srinagar, Kashmir.

Email: nihaqureshi01@gmail.com

22

Use of Tractor for Hauling Net

Hauling of net is a labour intensive task and requires people to pull out the net. To reduce this work-load, farmers of Haryana have devised a technique of using tractor to haul the net. The ends of the net are tied with the tractor and the tractor is driven. As the tractor moves, the net is hauled reducing the workload and labor. Farmers, who do not have easy availability of labor or timely accessibility of labor, have started using tractor for the purpose of hauling. It makes work easier and time efficient.

Practiced by: Farmers of Haryana

Documented by: Banti Debnath, M.F.Sc. Student, CIFE, Mumbai, Shri. Rishabh Sharma JNS Student, Mumbai and Dr. Arpita Sharma, Principal Scientist, ICAR- CIFE, Mumbai.

23

'Trapi': A Fishing Gear for Deep Water Fishing

It's a special type of fishing gear made up of bamboo strips. It has cylindrical shape and both the ends are opened. Bamboo strips are lined by special type of nylon wire and arranged densely so that the fishes cannot escape from the trap. It has only one window at the bottom where the fishes can be collected. As it is being mainly used in deep water with moderate water flow, height is somewhat higher than other enclosures. The basic advantage is that the fishes trapped in it can be kept under live conditions for several days as it allows a regular water flow, and the structure is durable enough and resistant to water movement too. It is generally placed at the bottom of the water body and not allowed to be displaced under strong water movement by fixing it at a suitable place on the water bed.

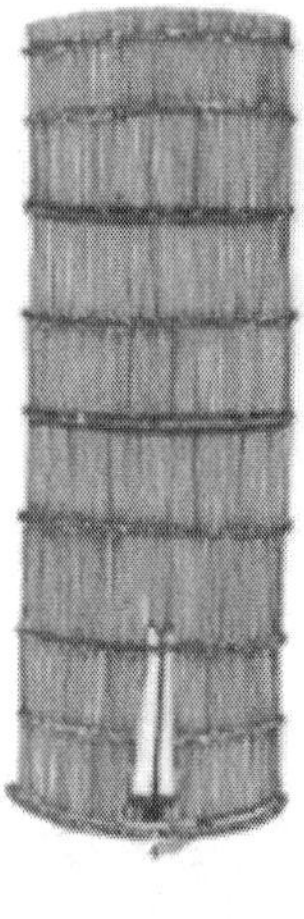

Practiced by: Jugnu, Fish farmer, Thane, Maharashtra

Documented by: Shri. Amitava Ghosh, M.F.Sc Student, CIFE, Mumbai

Email: amitava.skg@gmail.com

24

Use of Sand Bed for Maintaining Temperature of Alcohol Preserved Pituitary Glands Used for Induced Breeding

Pituitary glands are used for induced breeding of fishes. In general, pituitary glands are supplied to hatchery owners by the marketing agents. For successful breeding the quality of the pituitary gland is of paramount importance. That is why the glands are to be preserved properly. Generally, pituitary glands are preserved in alcohol, wherein this alcohol is changed at a specific interval of time. Temperature plays a crucial role while preserving the glands and it is necessary to maintain a cool temperature while storing. However, it is very difficult for hatchery managers to store in such controlled environment in farm site.

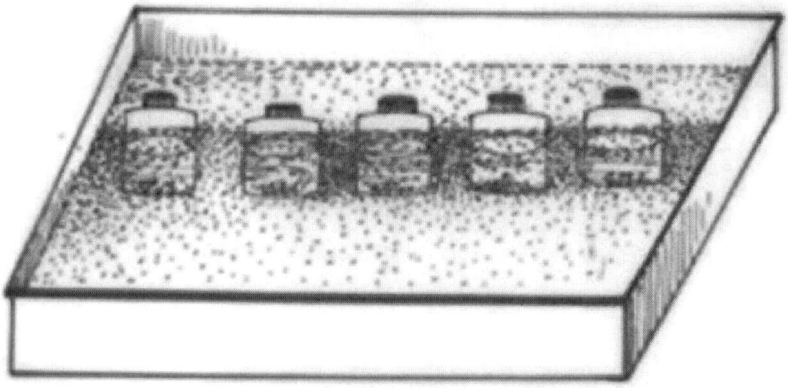

To solve such problems, hatchery owners in many parts of the country including North 24-Parganas district of West Bengal have developed a simple innovative idea of using sand bed for maintaining a congenial range of temperature for preservation of pituitary glands. A sand bed in a metal tray having a thickness of 4-5 cm is prepared where two third of the small bottles of alcohol preserved pituitary glands are burrowed. As bottles are placed underneath the sand bed, external radiation and temperature gets reduced and temperature can be maintained properly for a longer duration. Importance of this innovation is that it is cost effective and economical. This reduces the dependence of the farmers for the supply of the commercially available pituitary extract.

Practiced by: Shri. Babul Majumder, Hatchery owner, North-24 Parganas, West Bengal

Documented by: Shri. Amitava Ghosh, M.F.Sc Research Scholar, ICAR-CIFE, Mumbai

Email: amitava.skg@gmail.com

25

Floating Bed Concept: A New Innovation in Integrated Fish Farming

Water hyacinth which otherwise is thrown away as unwanted material has been utilized by farmers in an innovative integrated farming system. The floating bed prepared with the weeds (water hyacinth) are utilized for integrated farming of fish cum vegetables. The water hyacinth is gathered in one corner of the pond with the help of bamboo to form a bed of the size of 12 ft X 4 ft. Some portion of the weed is allowed to decompose on the embankment. The bed is regularly watered, and after a period of four days, a layer of decomposed water hyacinth is put over the bed. Hereafter, vegetable seeds, mainly spinach are sown over the bed. Seeds germinate there and grow on the floating bed, which reach harvestable size after two months. The remaining part of the pond is used for fish culture. This innovative technology helps in proper utilization of the weed and weed infested ponds for fish cum vegetable farming. This innovative use of weeds for fish cum vegetable culture has augmented the income of the fish farmers.

Practiced by: Shri.Mukunda Saikia, Borthakarabari village under Darrang District, Assam

Documented by: Shri.Apurba Bhuyan, Bhupen Kr. Daflari and Simanku Borah

26

Water Recirculatory System to Reduce Obnoxious Gases and to Maintain Dissolved Oxygen (DO) in Ponds

Re-circulatory system helps in reducing concentration of obnoxious gases and maintaining DO level in pond. So, aerators are not needed in ponds. However, the cost of setting up of a re-circulatory system is high. To solve such issues, the fish farmers of Tripura have developed an innovative method of maintaining the DO level in interconnected ponds. This system is practiced in 5 ponds which are located side by side and are interconnected through pipes at certain points. The pipes are placed at the bottom of the pond. To know the location of the pipe, bamboo poles are fixed over the point where the pipe is connected. This height of the bamboo pole is above the pond water level, and hence, is visible from outside. The point where the pipe is present can be easily figured out. Two ponds are connected by one pipe which draws water from one pond to the next pond. The water flows from first pond to second pond due to difference in pond depth. The second pond is then connected to the third pond by another pipe, and similarly the five ponds are connected.

The pipes are blocked by placing cloth and bricks. When there is a condition of less DO and fishes start dying or floating on the surface in one pond (the pond of concern), the water is drawn from the nearby pond. The pipe mouth is opened and thus water flows in from the other pond. This increases the dilution of the pond and hence increases the DO of the pond. This also facilitates mixing of the water due to the incoming water and hence obnoxious gases accumulated in the pond bottom are released.

Practiced by: Shri.Manindra Namo, Santibazar

Documented by: Shri.Tarikool Islam, B.F.Sc., College of Fisheries, Agartala, Tripura **Email:** tarikulcof@gmail.com

27

Application of Fermented Cow Dung and Paddy Straw

In carp culture, cattle dung is the most commonly used organic manure. Sustained zooplankton production is very important during the initial nursery stage of fish culture for which manuring is practiced using cattle dung. If the fertilization using cow dung is continued after stocking, when the fingerlings have other means of feeding, it results eutrophication causing stress to the fish. To avoid this, cow dung is applied after decomposing it for few days. For this, cow dung and paddy straw are kept in a polythene bag for 10 days so that it gets decomposed. By doing this, the cow dung becomes rich in harmless bacteria which degrade complex lingo- cellulosic compounds to simpler forms and in turn it promotes plankton growth.

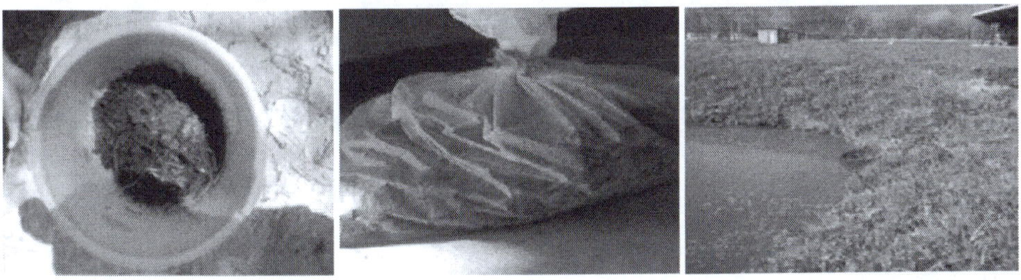

The decomposed cow dung is placed in a circular willow basket made of tightly woven twigs, and is placed submersed at one corner of the pond. This ensures moderate fertilization and support sufficient plankton growth in the fish pond.

Practiced by: Dr. Salman Chalkoo, Shopian, Kashmir

Documented by: Shri.Sheikh Imtaiz Qayoom, Sheikh Shafat Hussain, Syed Aalia Mushtaq, Iram Farooq and Sauliheen Qadri, Post Graduate Students, College of Fisheries Srinagar, Kashmir

Email: rizwana1309@gmail.com

28

Use of Modified Rickshaw in Fish and Related Commodity Transportation

Chakda is a three-wheeler transporting system popular in Gujarat. It requires less time for long distance transportation and has an ability for transportation on poor roads. In Veraval fish landing center of Gujarat, the transportation of fish material, oil for vessels and ice for fish preservation are mainly done through the chakda. This vehicle is made of a bike which is modified and brightly decorated. Bike with strong engines is preferred with different fuel acceptance capabilities like diesel and petrol. This vehicle is made locally.

The specification of this vehicle locally known as Chakda is 3150 x 1370 x 2800 (L X W X H) with a displacement of 391 cc for prteol/LPG/CNG and 395 cc for diesels. The GCW is 1000 kgs. Fuel tank is 14 litres and brakes are Fand R drum type.

Practiced by: Fishers of Veraval Taluka – Veraval, District - Junagadh, Gujarat

Documented by: Mr.Vikas, Village Bijethua Rajapur, Post – Mangrava, District : Sultanpur (U.P.) – 228161, **Email:** vkspathak20@gmail.com

29

Matsyajan: Mobile Insulated Fish Vending Unit

'Matsyajan' or mobile insulated fish vending unit is a scheme of Rashtriya krishi Vikas Yojana (RKVY) 2011-2012 with Fisheries department, West Bengal.

'Matsyajan' is a three wheeler cycle used as mobile unit for vending frozen/ fresh fish. It has also been fabricated to transport live fish. It has an insulated box (poly-ethylene) of 70 litre capacity, which has 4 separate containers for keeping fishes, two for live fish and two for fresh chilled/iced fish. Ice is used for freezing. Water is kept in two containers, which are used for carrying live fish. The containers are provided with a pipe to drain out melted ice. Mainly Indian Major Carp (rohu, catla, mrigal,) and exotic carp are stored for selling, both in live and fresh condition. Sometime live fishes like koi, singhi and magur are also kept for selling. The fish kept in frozen container remains in good condition for 2-3 days. The unit is fitted with a loudspeaker for advertisement of the products it is carrying init. The entire body of the unit is painted to display the products and their price. Light arrangement is also provided in the 'Matsyajan'. The unit has a solar system (battery, panel) to supply the required energy for audio system and for light. The unit also contain a Government approved weighing balance. The cost of the total unit is Rs. 50,000/-. which is given by RKVY. One time working capital for purchasing fish from wholesale market is also given from Central Fishermens' Co-operative Society, Purulia. The total scheme is monitored under the supervision of Fisheries Department, Govt. of West Bengal. The maintenance of the unit is done by the Block Fishery Extension Officer. Ice in ratio of 1:1, is used for lowering the temperature of fish.

Practiced by: Jitu Koibarta, P.O.: Bandoan, Purulia, West Bengal.

Documented by: Shri.Aniruddha Roy, M.F.Sc, Department of Fish Processing Technology, Faculty of Fishery Sciences, W.B.U.A.F.S.

Email: aniroyffsc@gmail.com

30

Utilization of Masmin by Lakshadweep Fishers (Chetlat Island)

Lakshadweep sea is rich in fishery resources such as the tuna, bill fishes, pelagic sharks etc. and other ornamental fishes. The estimated marine fishery potential in Lakshadweep waters is about 63,000 to 1,40,000 tonnes of various groups of fishes, whereas, the present annual production is around 10% of the potential. At Lakshadweep over 85% of the total landings constitute tunas, out of which 50% of the total tuna landings are used for masmin production, 20% for canning, and remaining 30% is consumed fresh. Masmin is a product prepared by traditional fishers for a long time in Lakshadweep. Masmin is utilized for preparing curry products like biryani, fried masmin etc and some other value added products. Among them, masmin sauce (mass sharkara) and masmin samosa (mass appam) are important due to their preference and utility. Tunas especially Skipjack Tuna (*Katsuwonus pelamis*) and Yellowfin tuna (*Thunnus albacores*) are used for making masmin.

Masmin sauce is known as mass sharkara in Lakshadweep islands. It is a by-product from masmin. In the cooking steps, boiling in seawater causes the release of tissue fluids along with some nutrients as exudates. This exudate is retained and after 3 to 4 cycles of masmin production, these exudates are mixed together and further processed. The process of heating without water and additives is continued till it becomes highly viscous fluid like sauce. This is packed in plastic or glass containers and marketed. This viscous fluid is called as 'masmin' sauce or 'mass sharkara'. The product has a shelf life of 4-6 months. This sauce is added to the fish curry as an excellent taste enhancer, and it is readily available in local markets of Lakshadweep. Mass Appam is a value added product made from masmin.

The shape of the product is like samosa which is available in restaurants and bakery having good demand from tourists. For making 1 kg Mass Appam, Ingredients required are : Masmin Chopped- 700 gm, Wheat Flour/ Maida- 500 gms, Chilli Powder- 30 gms, Onion Finely Chopped-50 gms, Grounded Coconut- 100 gms, salt as per requirement. Masmin is chopped into pieces and grounded into small flakes in a mixer grinder. Frying pan is heated and masmin flakes are also heated without oil for a few minutes and kept aside (Avoid charring). Chopped onion is fried in same frying pan till pale brown colour is obtained. Heated masmin is put in this along with chilli powder and salt for a few seconds and kept aside. The wheat flour/Maida is made into dough and shaped into thin round sheets like chapattis. The heated masmin with ingredients is stuffed inside thin maida sheets and kept in the mould and shaped as per requirement. The moulded form is deep fried in oil. Different kinds of moulds are used for this purpose.

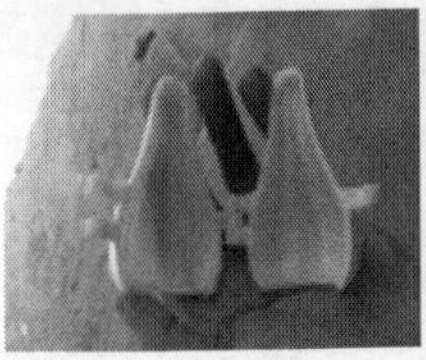
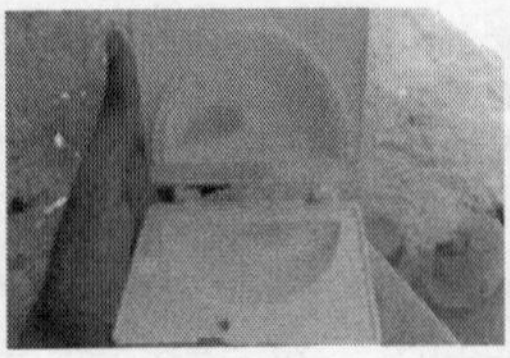

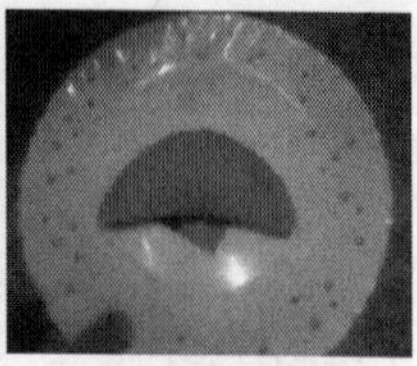

If the product has to be stored for a longer period then, onions and coconuts are to be avoided because it creates undesirable flavours in the product. The shelf life of the product is one month (without coconut) and 2 weeks (with coconut) Mass appam production is mainly done in home studs for restaurants and bakery, and has got good popularity among domiciles and tourists.

Practiced by: Shri.Dilkush.Ali, Manmel House, Chetlat Island, Lakshadweep

Documented by: Mr.Vishnu. R. Nair, M.F.Sc. Fish Processing Technology, KUFOS Panangad Kochin, Kerala.

Email: vishnuravi99@gmail.com

31

Catching of Crabs

Crab catching is a dangerous but interesting job and the technique used is unique. In this technique, one end of the string is tied to the small weight, making sure that the knot is secure. A bait is added (for bait mainly turbo and the cheliped of the crab are used) by tying it to the string near the weight.

The other end of the string is tied with the small piece of wood. The string is waded into the water (in the puddles, small creeks or crevices) and sinked with the bait and stand still. When the tension is felt in the string, the string is pulled out of the puddles gently and steadily until the crab is visible on the surface. Then a net is used to scoop the crab. The crab is grabbed from the net, holding it by the base of one of its legs while avoiding the claws. Then chelipeds of crab are tied along with their carapace with the help of a thin nylon/cotton string and kept in a bucket. Crabs can survive for several hours in the bucket.

Practiced by: Fishers of Veraval Taluka, Veraval, District: Junagadh, Gujarat

Documented by: Shri.Vikas, Village Bijethua Rajapur, Post: Mangrava, District: Sultanpur-228161 (Uttar Pradesh)

E mail: vkspathak20@gmail.com

32

Smoke Drying of Trapa Bipinosa (Water Chestnut) in a Specially Designed Klin

The fishers living along the Wular lake of Kashmir valley, harvest a huge catch of fish from the lake particularly during summer season. The fishers also get a good amount of water chestnut (Trapa bipinosa) from the lake which they smoke dry in a specially designed oven or kiln and sell it in local markets or to middlemen. This kiln is made up of locally available tree trunks.

The Kiln consists of a framework made up of wood raised on a platform from the ground. The roof of the kiln is kept covered with gunny bags so that smoke escapes slowly. The wet trapa are kept on the frame work and fire is ignited in the kiln, and the whole kiln is covered with the gunny bags on all sides. After few hours, the gunny bags are removed from the top and front of the frame, and smoke is allowed to escape completely. The dried chestnuts are spread on a LDPE sheets and then sold in the market. The external cover of the water chestnuts is also used as a feed in carp culture.

Practiced by: Fishers of Wular lake, Laherwal pora , Bandipora district

Documented by: Neha W. Qureshi, Ph.D. Scholar and Dr. M.Krishnan, Head FEES Division and Principal Scientist, CIFE, Mumbai

Email: nihaqureshi01@gmail.com

33

Short and Sharp Aerator

One of the most common problems prevalent in aquaculture is low dissolved oxygen which causes high fish mortality. Marginal farmers can not afford to buy aerator for their farm. To solve such issue, farmers of Tamil Nadu have devised an aeration equipment consisting of three pieces of PVC pipe, plastic empty can (20 lit. capacity) and a half HP motor.

The aerator structure is fixed inside the pond, generally in the centre of the pond. The main objective behind this is to overcome power shortage. This equipment uses single phase power which helps the aerator to run continuously, thus reducing the problem of low dissolved oxygen in pond system.

Moreover it does not cost much as it can be developed within Rs. 5000. It can work as other aerator (paddle wheel, bubble etc.) and is found adequate for an area of 1000 m2. This can be used in overhead tank, small pond, nursery pond etc. This type of aerator can be used by small scale and large scale farmers.

Practiced by: Suresh Raja, Tamil Nadu,

Email: himalayanaquatech@yahoo.co.in

Documented by: Shri.Balaji Guguloth, M.Sc. Student, FC & RI, Tuticorin
Email: gbalu002@gmail.com

34

Use of Grass Mat in Aquarium

Ornamental fishes like dollar and zebra do not exhibit parental care after laying eggs. In fact brooder fishes treat eggs as good source of nutrients and feed on them. Hence use of grass mat in aquarium is found to be useful as they trap eggs and do not allow brooders to feed on them. After laying eggs, brooders are removed and kept in another aquarium.

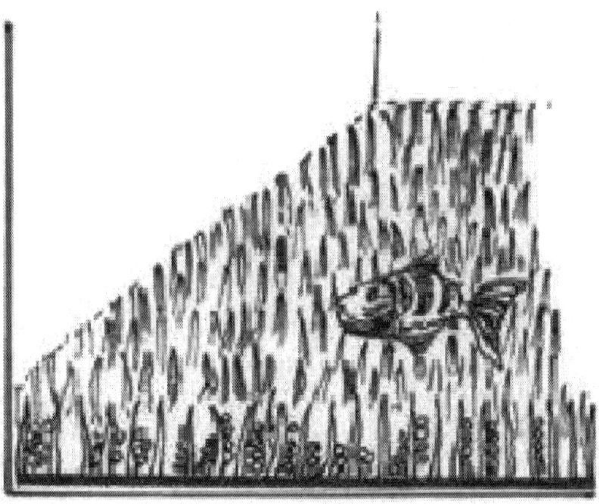

Whereas eggs are reared in the same aquarium trapped in grass mat till it hatches. After hatching, hatchlings are collected and kept in another aquarium and reared. Grass mat is stabilized in aquarium with use of PVC pipes that are filled with sand or mud and sealed. These pipes act as weights while providing a hygienic aquarium environment. Grass mat also act as bio filter and reduces ammonia in aquarium water, thus provides a healthy environment to the fish.

Practiced by: Dr. S.G.S. Zaidi, Technical Officer, ICAR-CIFE, Mumbai

Email: sgszaidi@hotmail.com

Documented by: Shivta Kureel, Ph.D Scholar, ICAR-CIFE, Mumbai

Email: kureelshivta@yahoo.com

35

Raking of Pond Bottom Through Hand-Tractor and Metal Barbed Structure

After dewatering pond water, the pond bed is dried and raked to get rid of obnoxious gases and weeds. There are different ways of raking. In West Bengal, indigenously built structure known as "Hara" is used for raking pond bottom for removal of obnoxious gasses.

Generally "Hara" is made up of cluster of stones, bamboo sticks, barbs which are bundled up and finally made a heavy structure remaining anchored in the bottom while it pulls out. Both the ends of the structure is tied with ropes and pulled manually by fishers. But this practice is very labour intensive, time consuming and costly too.

To get rid of this problem a farmer of Nadia district in West Bengal has replaced the "Hara" with galvanized-iron made barbed structure and for pulling the structure, hand-tractors are being used. Here two ropes which are tied with both ends of the structure are fitted with back side of two hand-tractors, which are placed on both sides of the pond embankment. Both the hand-tractors start moving with equal relative speed through the embankments, enabling the movement of iron made structure through the pond bottom, resulting raking of pond bottom uniformly.

Practiced by: Shri.Nirapada Ghosh, Fish farmer, Nadia, West Bengal
Documented by: Mr.Amitava Ghosh, M.F.Sc. Student, ICAR-CIFE, Mumbai
Email: amitava.skg@gmail.com

36

Irrigation Free Horti-Cum-Fish Culture Practice in Rocky Areas

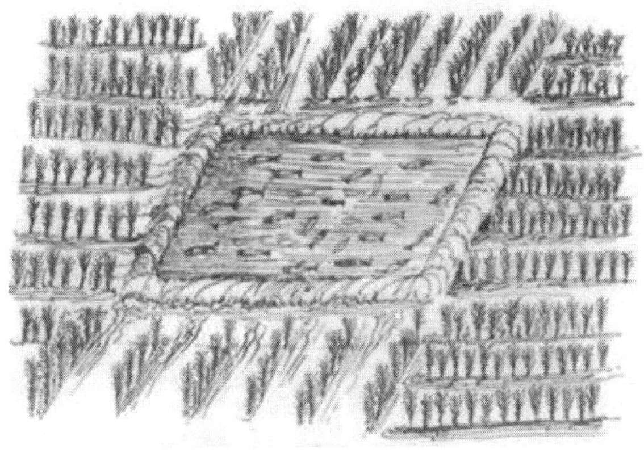

High permeability is the characteristic feature of rocky soils which makes fish culture and agri-irrigation very problematic due to less water retention in ponds, beels and canals. But Farmers of Kharkadh village of Maharashtra have developed an innovative practice to get rid of such problem in high permeable soil.

They have dug a pond in upland of a rocky soil area which is filled with rain water and remained filled till summer months. Due to high permeability, water leaches out from the pond and gets absorbed in adjacent low land areas of the pond. Different horticultural crops have been planted in the adjacent low land areas of the pond which get irrigated due to seepage of water from the pond automatically. It is very interesting to note that no additional irrigation is required for the plants and simultaneously the pond is being used for fish culture.

Practiced by: Farmers of Kharkhad, village, Maharastra

Documented by: Mr.Amitava Ghosh, Mr.Abhijeet Naohate, M.F.Sc. Studnets and Shri.Aritra Bera, Ph.D. Research Scholar, ICAR-CIFE, Mumbai

Email: amitava.skg@gmail.com

37

Use of Perforated PVC Pipes to Keep Brooder Prawns

When male and female brooder prawns are kept together they fight amongst themselves and cause injury to each other. It leads to multiple injury, loss/ damage of limbs and even mortalit y. To avoid such problem farmers use perforated PVC pipes to keep male and female separately in the breeding tanks/ tubs. The perforated PVC pipes with multiple holes are kept in breeding tanks/ tubs and the male and female prawn brooders are kept separately in these pipes. These perforated

PVC pipes are of about three to four inch of diameter and covered with a net on one side. After keeping the brooder prawns inside this pipe, other opening is also covered using net so that brooders cannot come out. Holes in PVC pipes and nets on both side allows proper circulation of feed and water. By this way the brooders remain separately and cannot fight. This practice helps brooders to be safe and healthy.

Practiced by: Farmers of Kakinada Farm, Andhra Pradesh

Documented by: Shivta Kureel, Ph.D Scholar and Mr S. K. Sharma, Technical Officer, ICAR-CIFE, Mumbai

Email: kureelshivta@yahoo.com | **Email:** sksharma.icar@gmail.com

38

Removal of Adhesiveness of Fish Eggs by Using 'Alamati' (Laterite Soil)

Seed production through hatchery technology is very common and lucrative venture. For successful hatchery operation there are several factors which play a major role in sustainable seed production. One of such factor is the adhesive nature of certain fish eggs. Generally, common carp eggs are adhesive in nature.

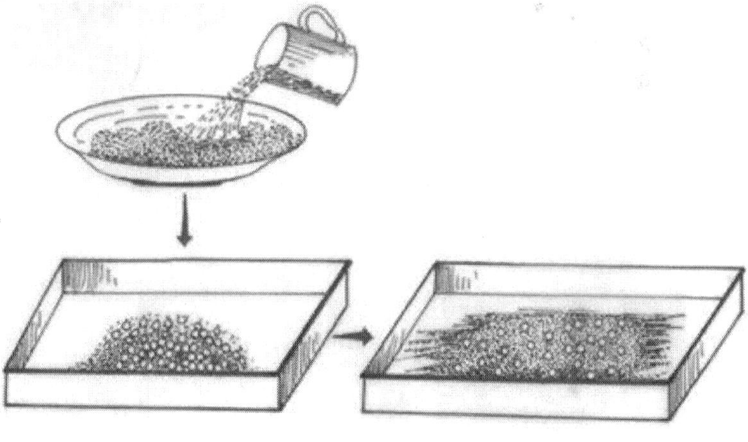

During the course of induced breeding fish eggs are stripped out and allowed to mix with sperm. But for this purpose, the eggs have to be free from adhesiveness otherwise mixing with sperm will not be proper. For removing the adhesiveness use of milk powder and buffalo's milk is very common, but these are expensive. As a low cost alternative, a farmer in North-24 Parganas of West Bengal has been practising a special type of laterite soil solution locally known as 'alamati' for removal of eggs adhesiveness. As this soil has rich iron content, it helps in absorbing the adhesive mucus of fish eggs. It can also be used for other fish with such eggs.

Practiced by: Shri.Babul Majumder, Hatchery Owner, North-24 Parganas, West Bengal

Documented by: Shri. Amitava Ghosh, M.F.Sc. Student, ICAR-CIFE, Mumbai

Email: amitava.skg@gmail.com

39

Use of Thorny Shrubs to Save Fishes and Crustacean

Thorny shrubs are placed in water and also around the water body in order to save fishes and crustaceans to prevent poaching. Shrubs are planted in the embankments of pond to prevent entry of poachers in the pond. A secret passage is left open known only to the farm owner that is used during harvesting. When thorny shrubs are placed in pond water, a rope is tied to each shrub with help of which all the shrubs are pulled out to remove it before harvesting. Because of these thorny shrubs poachers are not able to enter the pond, hence fishes and crustaceans are saved.

Practiced by: Farmers of Andhra Pradesh

Documented by: Shivta Kureel, Ph.D Scholar and Shri. S. K. Sharma, Technical Officer, ICAR-CIFE, Mumbai

Email: kureelshivta@yahoo.com | **Email:** sksharma.icar@gmail.com

40

Durable Scoop Net Made of Old Aeration Tube

Scoop nets available in market are made of iron rod covered with plastic mould extending upto the handle and a fine net attached to it. These scoop nets are not durable and get damaged within 3 to 4 months. A modification has been made over it where old aeration pipe is used to cover the iron frame.

The PVC pipe fixed as handle can also be covered with such aeration pipe. This enhances the durability for more than 8 months of the scoop net. It is also convenient and easy to handle.

Practiced by: Dr. S.G.S. Zaidi, Technical Officer, ICAR-CIFE, Mumbai

Email: sgszaidi@hotmail.com

Documented by: Shivta Kureel, Ph.D Scholar, ICAR-CIFE, Mumbai

Email: kureelshivta@yahoo.com

41

Use of Empty Canister (Plastic Drums) to Make Fish Cages

Cage culture is one of the popular methods of fish culture specially in open water bodies. But the cost of a cage is huge when purchased from the market. As a solution to that, farmers developed a design of cage using locally available materials.

The structure of the cage is made by placing four empty canisters or plastic drums in four corners and joined by bamboo in order to form a square frame.

Nets of desired mesh size are hanged from the bamboo frame enclosing from all the sides which give the structure of a cage. Fishes are reared inside these structure in open water, which enables better environment for growth and immunity of fishes.

Practiced by: Farmers of Bihar

Documented by: Shweta Kumari, M.Sc. Student, ICAR-CIFE, Mumbai

Email: shwetaniki88@gmail.com

42

Use of Turmeric and Henna Paste by Prawn Peelers to Reduce Hand Injuries

Prawn peeling is a task which is carried out in processing units as well as in the market. Women who are involved in prawn peeling suffer from a number of injuries on their hands and palms due to the pricking of rostrum of prawns and continuous exposure to the ice. To get relief to the palms, fisher women apply henna (*Lawsonia inermis*) and turmeric.

(Curcuma longa) paste at night on their palms and remove the paste in the morning.

Practiced by: Prawn Peelers of Versova, Mumbai

Documented by: Banti Debnath, Ph.D. Scholar, ICAR-CIFE, Mumbai, and Dr. Arpita Sharma, Principal Scientist, CIFE, Mumbai

43

Tide-Fed Canal Based Aquaculture System in Mangrove Backwaters

In India, brackishwater aquaculture depends mainly on shrimp culture due to its commercially viability. This technology cannot be used for fin fishes and crabs because of high input cost. In case of mud crab culture, due to the problem of cannibalism the survival rate is very low, thus the commercial cultivation of crabs is limited.

Inlet Sluice

Outline Sluice

Integrated mangrove tide-fed aqua culture system is an innovative model to integrate forestation of mangroves with cultivation of fish, prawns and crabs in the symbiotic manner. It act as a source of livelihood for the coastal community as well as biological barrier against natural calamities. This system plays an important role in the coastal zone management. The model has been designed by keeping in mind the technology capable of reducing the production cost, the utilization of natural feed (live feeds along with candidate species), energy efficiency of tidal flow used (no electricity needed), less labour intensive due to semi-automatic, hideouts in the farming system to reduce the cannibalism and thus maintenance of survival and the biomass, protection from natural calamities like tsunami and cyclone for the people and the farm, socially acceptable by the stakeholders, replicable model for the other region of the country.

One of the pond at Puliyanthurai village, Sirkali Taluk, Nagapattinam district with an area of 0.70 hectare has been taken for demonstration of the idea to integrate mangroves with aquaculture. The canal has been dug in zigzag manner throughout the pond. On one end the canal is connected to the adjacent backwater through inlet sluice fitted with stainless steel mesh and a swing door arranged in such a way that it will allow only water to flow in during high tide and it will automatically close when water starts receding. Minimum water level of one meter is being maintained in the pond. The other end of the zigzag canal is connected to the backwater by the outlet sluice gate which is also fitted with stainless steel mesh and a swing door, but this swing door works in the opposite manner. During high tide it remains closed and starts draining the water during the receding tide into the backwater. So the water enters through one end and drains through the opposite end making the water quality in canal very much suitable for aquaculture. All along the edge of the canals, several species of mangroves such as *Rhizophora sp, Avicenia sp.* and *Sesuvium sp.* have been planted to avoid soil erosion. For the plantation of mangroves and halophytes, the natural zonation pattern is followed for better growth and development. This is for the first time that mangroves are artificially planted and grown for the purpose of aquaculture wherein both intake and drainage is done by tidal flow. In a day, 40-70% of water is exchanged for two times. Hence there is no need of using power (electricity/diesel) for pumping/aeration. Only at the time of harvest, pumps are used to drain the bottom water. Hence it is almost a zero energy model. The stilt root of *Rhizophora sp, pneumatophores of Avicenia sp.* serve as the effective natural hide out for mud crabs. The mangroves serve as the biological barrier against the natural calamities like flood, cyclone, tsunami etc. The inlet and outlet sluice gates are automatically operated by tidal pressure, so there is no need of labour for the water quality management.

Practiced by: Famers of Tamilnadu

Documented by: Shri.Ashwini Kumar Singh, PG Scholar, Department of Fisheries Extension, Fisheries College and Research Institute, (TNFU), Tuticorin-628008 Tamilnadu

Email: ashwinifishco@gmail.com

44

Homemade Sinkers by Tribal Fishers of Nicobar

Most of the fishing gears such as gillnet, cast net, long line are fitted with sinkers and need to procure while making a net. To reduce the cost of sinkers, a fisher from Nicobar island fabricates different types of sinkers using locally available materials. Different shape and size of sinker are made by using clay and iron mould of cone or cylindrical shape. Such use of localy available materials reduces the cost of net. The process is shown in the following photographs:

Melting of lead

Removal of the sinker from the mould

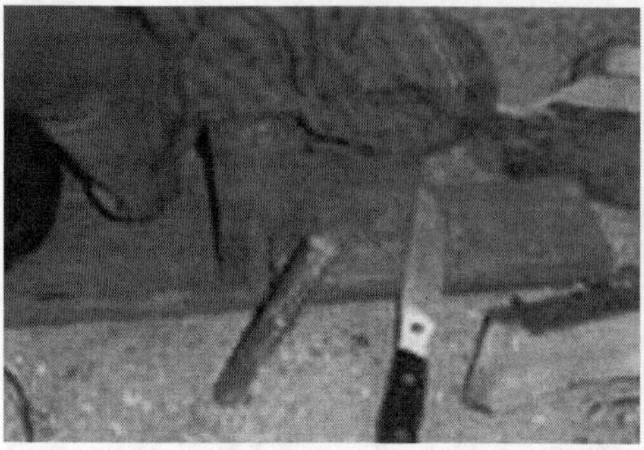

Fixing of thin iron rod for the hole

Clay shaped mould structure

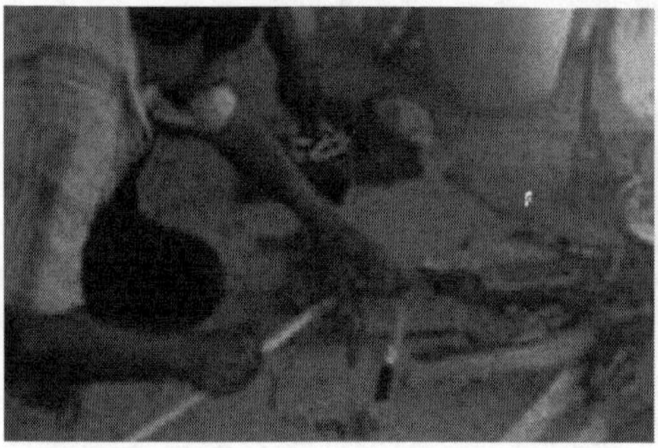

Pouring of lead in the iron mould

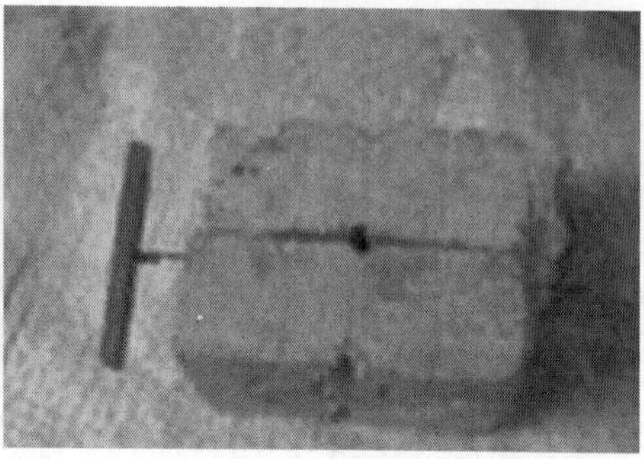

Clay shaped mould

Sinkers of different shapes made it in the wet land mould by using paper

Making of required shape hole in a wet land

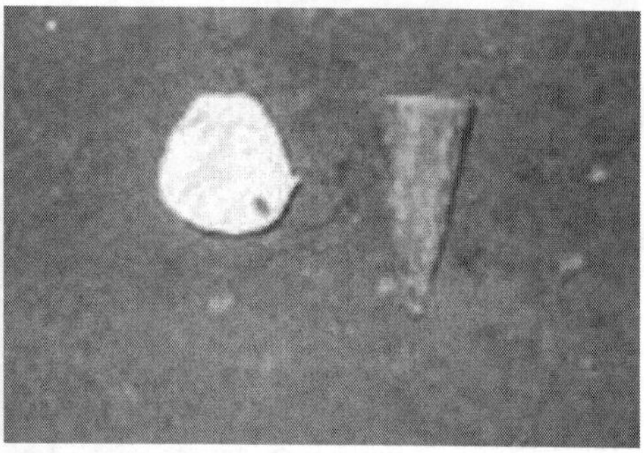

Fixing a paper in the hole

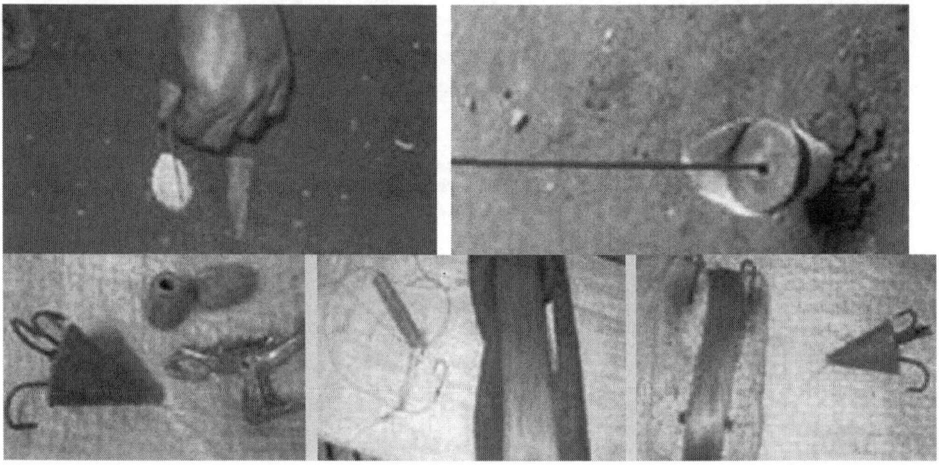

Different types of sinkers made through technique

Practiced by: Shri.Nasir Hussain, Teetop village, Car Nicobar, Nicobars District, Andaman and Nicobar Island (India)

Documented by: Shri. Vikas Kumar, Junior Research Fellow, Department of Fish Processing Technology, Fisheries College & Research Institute, Thoothukudi- 628008 (TN), India.

Email: vikasfpt@gmail.com

Dr. Subhash Chand, Senior Scientist, Agricultural Economics, Central Agricultural Research Institute, Port Blair, India- 744105

Email: scchand46@gmail.com

45

Indigenous Snorkeling Kit

The saying "Necessity is the mother of invention" is well reflected in fishing crafts and gears invented by these tribal fishers. A fisher from Nicobar island fabricates snorkeling kit using locally available materials such as cycle tube, broken glass and aluminum or PVC pipe.

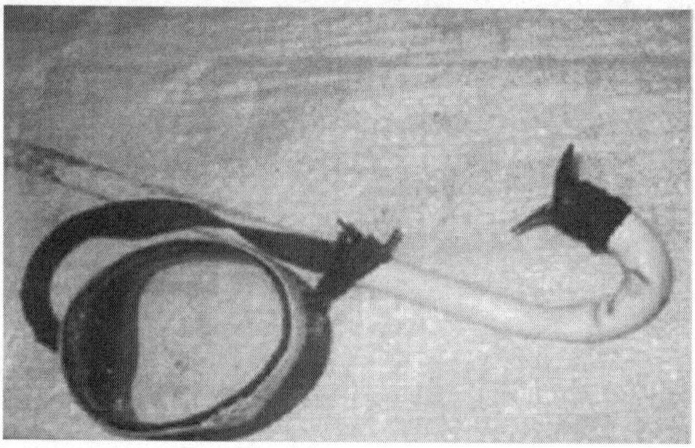

Practiced by: Shri.Nasir Hussain, Teetop village, Car Nicobar, Nicobars District, Andaman and Nicobar Island (India)

Documented by: Shri.Vikas Kumar, Junior Research Fellow, Department of Fish Processing Technology, Fisheries College & Research Institute, Thoothukudi- 628008 (TN), India.

Email: vikasfpt@gmail.com, Mob: 07305260934.

Dr. Subhash Chand, Senior Scientist, Agricultural Economics, Central Agricultural Research Institute, Port Blair, India- 744105

Email: scchand46@gmail.com

46

Use of Wind Mill for Preservation of Fish Catch in Fish Holds Onboard Vessel

Photo 1. A Modern Fishing Vessel with wind mill and solar panel

Photo 2. Refrigerated fish hold

One of the major problems of marine fishers is non-availability of ice for onboard preservation and storage of fish catch. This problem is intensified by non-availability of electricity for production of ice.

In view of the same, electricity is produced onboard vessel by wind mill and solar panels. The electricity thus produced, is used for keeping the fish cold at less than 00 C.

Special stainless steel plates are used for construction of insulated and refrigerated fish hold onboard vessel and same is supplied with electricity produced from wind mill and solar panel.

Prime advantage of this is that diesel consumption (saving of Rs. 7000/-per month), and ice consumption (saving of Rs 8000/- per month) is significantly reduced as fish catches are handled and stored hygienically. Thus, protein-rich valuable fishes are not being wasted and earnings of fishers also increased many folds.

Practiced by: Shri.Vithal Bhalekar, Post: Veldur, Taluka Guhagar, Dist: Ratnagiri

Documented by: Shri.Chandresh Haldankar, Shri.Amit Kokate, Miss Pooja Bhalekar, College of Fisheries, Shirgaon, Ratnagiri, Maharahstra

Email: chandreshhaldankar@gmail.com

47

Refugee Lata (*Mikania Cordata*) Plant Extract to Control Epizootic Ulcerative Syndrome Disease

Epizootic ulcerative syndrome (EUS) is caused by mainly fungus, Aphanomyces invadans. The disease occurs in low temperature and during heavy rainfall in tropical and sub-tropical waters. Till now there are very few methods or chemotherapeutic agents available to control EUS.

EUS occurs due to poor water quality management in the pond. In winters as water temperature is low, EUS spreads very easily. To combat such menace, farmers of Tripura have developed an effective method using locally available plants and herbs. The fishermen use Refugee lata (Mikania cordata) found easily in the rural areas of Tripura, and green turmeric juice together with ash for treatment of EUS in their ponds. It is practiced before the onset of winter and monsoon season. The basic requirement for control of EUS is to make the water alkaline. Ash makes the water alkaline. Thus, the application leaf extracts, ash and turmeric, which has antimicrobial action, has been reported to control this syndrome to a great extent.

Practiced b : Nikhil Mazumdar, Chief Executive Fisheries Officer, ADC, Khumlung, Udaipur, Tripura

Documented by: Luigi Pangsang M Sangma, B.F.Sc, College of Fisheries, Lembucherra, West Tripura

Email: lsnosang@gmail.com

48

Engine Used to Pull Fishing Boat to Shore

It has always been a problem and labour intensive to pull fishing boats to shore area when it is not operating. To reduce such labour, fishers pull their boats to shore with the help of engine. A rope/ metal wire is tied to the boat and connected directly to engine, which reduces a lot of time and man power. Since it is operated by kerosene, the process is economical and practically viable.

Practiced by: Local fisherman of Jhaleshwar (Veraval), Gujarat Documented by : Shyam Kumar, Muzaffarpur, Bihar 842002

E mail: shmkumar265@gmail.com

49

Semi-Automatic Cage Mariculture

Offshore aquaculture is a challenging task for farmers as feeding and harvesting in mariculture is carried out in sea. Weather conditions are not equal at all the time. For this, 'Semi-automatic cage for mariculture' was developed.

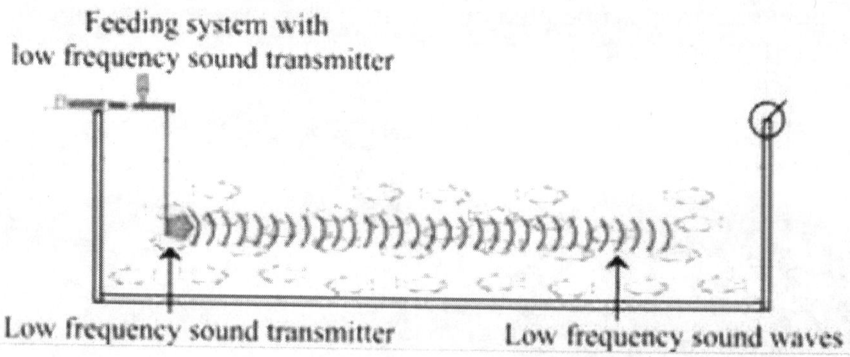

Automation in mariculture with use of low frequency sound for attracting fishes towards food, increase profit and quality, reduction of risk and labour. The machine can dispense a feed 8 times in a day according to requirement. In addition to all this, feeding system has low frequency sound transmitter which attracts the fishes towards feed dispensing point of the machine. Apart from automatic feeding system, cage has its semi-automatic fish harvesting system.

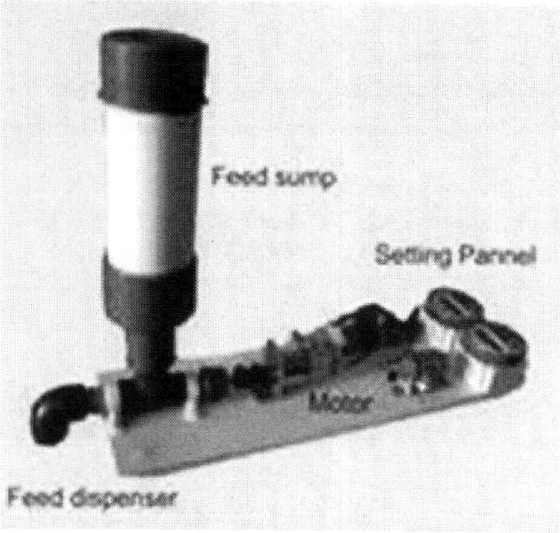

Documented by: Suyash Satyawan Patil, Shirgaon, Ganeshwadi, Tal. Dist. Ratnagiri.

Email: suyashfish@gmail.com

50

Domestic Cat Fish Egg Hatching Unit

Hatchery operation is one of the most important aspect of aquaculture as maximum mortality occurs during this period. The cost involvement is also more for maintaining the hatchery. Most of the hatcheries use several means to maintain water quality and survivality. As a way to reduce the cost of hatchery operation plastic tubs are being used for catfish hactcery. In this method, fertilized eggs are spread on the plastic tubs with inner chamber made up of plastic net (mosquito net). After hatching, spawns come out from the eggs and pass out from the inner net chamber to the main tank.

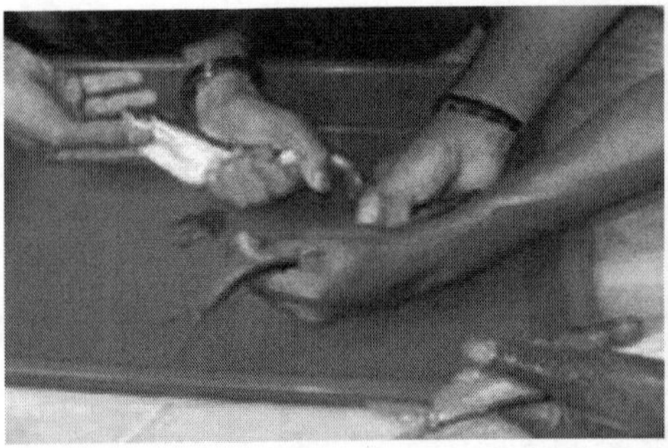

This helps to remove the egg shells easily and the water in the tub remains clean. The hatchlings of pabda do not need aeration as clean water contain DO of around 5ppm. Larvae can survive well in low DO as well as in comparatively derelict water, thus artificial aeration is not used for the purpose of rearing larvae of pabda. Advantage of this model is that the larval rearing unit can be set in any place without any specialized setup.

Practiced by: Pradyut Biswas, Lembucherra, Tripura (W) - 799210

Documented by: Himadri Saha and Asem Lembika Devi, College of Fisheries, CAU, Lembucherra, Tripura (W)-799210

Email: sahacofcau@gmail.com

51

Pot in Pot Cooling Device

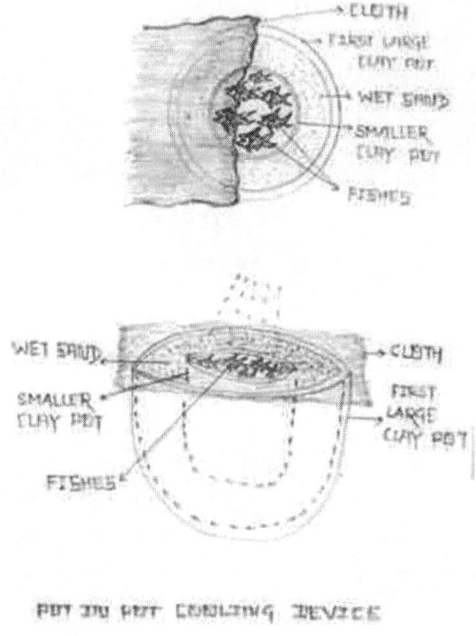

POT IN POT COOLING DEVICE

The pot in pot is a cooling device used to keep the harvested fish cool by using evaporative cooling mechanism. The device is utilized during selling of live fish and for transportation to shorter distances. This in turn fetch better prices. The technology includes two different sized pots. The smaller one is kept inside the larger pot while the space in between them is filled with wet sand and the top is covered with a wet cloth. When the water evaporates, it pulls the heat out, making the inner side cold. The advantages of using this technology include longer fish life and freshness in storage, easy manufacturing, easy availability of materials, cost effectiveness and easy usage.

Practiced by: Mohammed Bah Abba, Northern Nigeria

Documented by: Riya kumari and Vikas Kumar Sahu, B.F.Sc. II year, College of Fisheries, Kawardha

52

Sugarcane Bagasse for Ground Water Filtration

To avoid heavy metals in the culture ponds, farmers are utilizing sugarcane bagasse for producing activated carbon (AC) for groundwater. Toxic heavy metals from ground water are removed by adsorption of ions and the separation of sediments from water before releasing it into the ponds. Farmers collect sugarcane bagasse from local sugarcane juice shops. After drying for 4 days, it isput in feeder channel near bore well.

When water passes through the bagasse, heavy metals and iron content are absorbed and clear water passes towards pond inlet.

Practiced by: MSR Aquafarm, Dist. Cuddalore.,Tamilnadu

Documented by: Shri.Nitesh Kadtan, M.F.Sc. Student, ICAR-CIFE, Mumbai

Email: ans32006@yahoo.com

53

Fermented Palm Tree Water as Organic Fertilizer

Fermented water of palm tree have high percentage of carbon and acetic acid and it is a cheap source of organic fertilizer. This fermented water is being used as organic fertilizer in farm by shrimp farmer.

They use following dose – 5lit/ha along with wheat flour @ 10kg/ha, jaggery @ 10 kg/ha, rice powder – 15kg/ha and yeast @ 100gm/ha to fulfill carbon requirement of pond.

Practiced by: MSR Aquafarm, Dist. Cuddalore.,Tamilnadu

Documented by: Shri.Nitesh Kadtan, M.F.Sc. Student, ICAR-CIFE, Mumbai

Email: ans32006@yahoo.com

54

Removing of Fish Skin by Using Salt

It is very labour intensive to remove the skin of fish for dressing. As a solution to that, farmers use salt for removal of the skin. After cleaning the fish, 1 kg of large size fish is immersed in 3 kg of salt in a container for about 1 hour which helps in loosening the skin of the fish. After 1 hour, farmers slightly cut the fish skin with the help of knife and pull the skin by hand resulting easy removal of the fish skin.

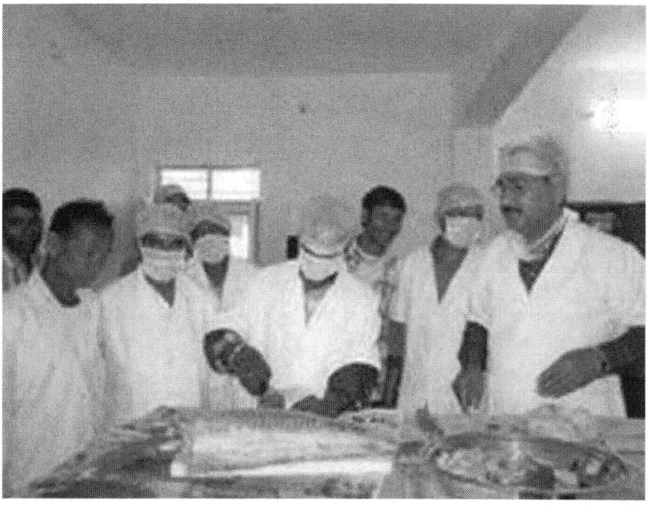

Practiced by: Fishers of Kerala, Palakkad

Documented by: Minal Sheshrao Wagde M.F.Sc. Student, CoF, MPUAT Udaipur, Rajasthan

Email: minalwagde@gmail.com

55

Crib Method for Manuring Pond

A compost crib constructed using wooden sticks at one or more sides of the pond. It helps fertilize the water gradually. The manure in the crib requires frequent turning to facilitate release of nutrients. It can be achieved by building a compost crib of 1 m radius in the corner of the pond for 100m2 pond water surface and stacking the organic waste materials in alternating layers under the water surface.

The upright sticks of the compost crib should be about 15 cm apart, so fish can enter and feed on worms and other food stuff. The compost crib ingredients should be mixed every day in order to allow nutrients to flow out. For a 100m2 pond about 50–60kg of organic matter is required to start with, followed by weekly doses of 8–10kg organic matter.

Practiced by: Talbindersingh, Rauni tehsil-Bharpurgarh, P.O, Amloh, Fathegarh Sahib, Punjab

Documented by: Sudhanshu Raman, BFSc Scholar Guru Angad Dev and Veterinary and Animal Science University, Ludhiana, Punjab

Email: sudhnshu147@gmail.com

56

Scissors for Dressing Fish

Traditionally fish sellers/vendors of the country use traditional cutter or knife for dressing fishes which involves more time and risks. In order to avoid these constraints, fishers use scissors of different sizes for the purpose of dressing fishes. They mainly use scissors for removing fins, gills and scales. Scissors are safer option and allow fishers to experiment with different designs.

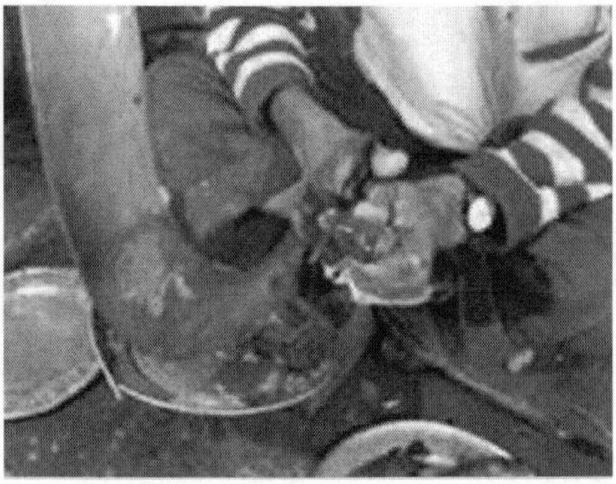

Practiced by: Fishers of Agartala, Tripura (W) - 799005.

Documented by: Himadri Saha, Assistant Professor, College of Fisheries (CAU), Lembucherra - 799210, Tripura

Email: sahacofcau@gmail.com

57

Chicken Viscera for Feeding Carnivorous Fish

Feeding of carnivorous fishes require feed containing high percentage of animal protein. Commercially available feeds are very expensive. As an alternative to that, fisher use viscera from slaughter centre (mainly chicken viscera) to feed the fish.

After collection of the viscera, it is processed properly and boiled. The boiled viscera are grinded/pasted before mixing with other ingredients like rice bran, wheat bran and mustard oil cake. Use of such slaughter wastes reduces the cost of fish production to a great extent.

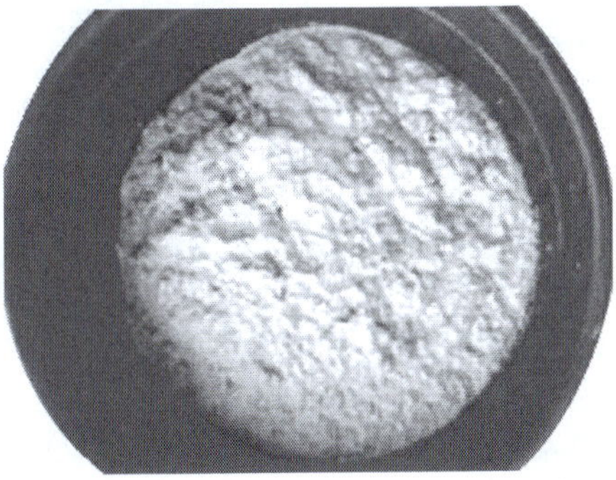

Practiced by: Himadri Saha, Assistant Professor, College of Fisheries (CAU), Lembucherra-799210, Tripura. Tel: 09774325853,

Email: sahacofcau@gmail.com

Documented by: Himadri Saha, Assistant Professor, College of Fisheries (CAU), Lembucherra-799210, Tripura.

Email: sahacofcau@gmail.com

58

In-Situ Feeding of Fish Larvae in Rearing Tanks

Every fish in early life stages can feed on the infusoria (protozoa). Culture of infusoria is quite easy as compared to other live organisms, however their collection is a bit difficult. To overcome this problem, banana peels are used. The peel is kept in 50litre of water and fish is stocked in the same tank (practiced with Zebra fish). The banana peel is changed once in two days to maintain the transparency of the water. Water exchange is carried out whenever the water became turbid.

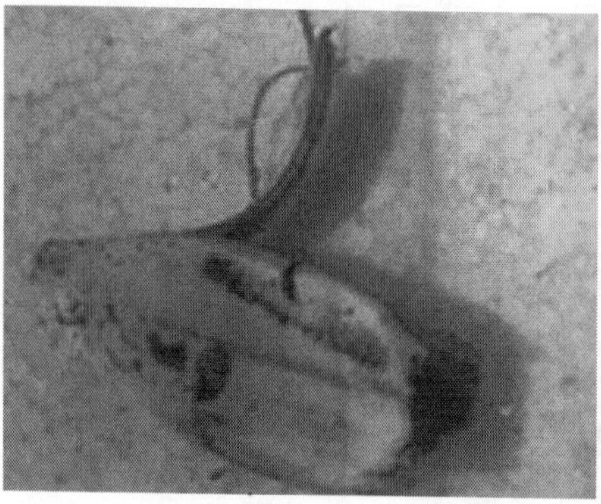

Practiced by: Mrs. Reba Saha and Dr. Himangshu Bhushan Saha, Mamani Aqua Research Centre, Abhoynagar, Agartala, Tripura (W)-799005.

Documented By: Himadri Saha, Assistant Professor, College of Fisheries (CAU), Lembucherra-799210, Tripura.

Email: sahacofcau@gmail.com

59

Stunted Yearling as Stocking Material for Fish Farming

As Indian major carps are known to possess faster growth rate in second year. The aqua culturists have taken advantage of this trait and started to use one year old stunted fingerlings as stocking material for grow out culture. Stunted yearlings are better stocking material for culture because of their higher survival rate as they are less vulnerable to predation, diseases and are more tolerant to environmental fluctuation; require less time to reach marketable size and also eliminate the problem of the seasonality and availability of carp seed for culture. The fingerlings are gorwn in a crowded condition for a year which make them stunted and after that suddenly they are provided with congenial environment which gives better growth compared to normal yearlings.

Practiced by: Prasit Das , S/o Pramoth Kumar Das, Ainri, P.O. Anantpur Block Soro, District : Balasore (Odisha) Pin: 756 046

Documented by: Ambika Prasad Nayak, Prof. D.Satapathy and Prof. (Mrs.) S.K.Mishra P.G.Department of Aquaculture, College of Fisheries (OUAT) Rangeilunda, Berhampur - 760 007 Odisha

Email: ambikaqua@yahoo.com

60

Low Cost Structures as an Alternative for Cement Tanks in Ornamental Fish Rearing Units

Low cost structures made with use of old cement bags filled with waste materials which are not easily degraded in nature and can be made into any size and shape, as and when required. Cement bags filled with materials such as sand, stones, plastic waste can be effectively used by putting lining of low cost polythene film inside it. Structures can be made of various capacities as per requirement.

Practiced by: Shri. Dinesh Agashe, At. Post. Karla, Ratnagiri

Documented by: Girija G. Phadke, PhD Scholar, Department of Fish Processing Technology, College of Fisheries - Technology Wing, Hoige Bazar, Mangalore – 575 001 Shrikrishna P. Hotekar, Senior Research Fellow, College of Fisheries, Shirgaon, Ratnagiri, Pin: 415 629 and Faisal R. Sofi, Kawkabul Saba, PhD Scholar, Faculty of Fisheries, SKUAST-K, J&K, Pin: 190 015

Email: girija_cof@yahoo.com

61

Strengthening Embankment by Bamboo Structure

This innovative structure helps in preventing soil erosion. Bamboos are commonly found and can be obtained easily. The cost of bamboo is cheap and in places where area is large it is grown near the ponds. A series of split bamboos are arranged horizontally with little or no space in-between. Some split bamboos are stuck vertically from the edge of the pond to the bottom to hold the structure together giving stability to it. This structure can be replaced after every 3-4 years.

Practiced by: Sunshine Shabong, Khatarshnong, Laitkroh, East Khasi Hills

Documented by: Daphishisha Kharhunai, V. Summinlun, Samchina Balsri G Momin, John H Kumar, Eborlang Kharnaior, Mawlai Phudmuri, Shillong - 793008

Email: lunvualnam@gmail.com

62

Traditional Method of Crab Fattening

Mudcrab fishing is carried out by using different methods like gillnetting, traps and handpicking. A fisherman carried crab fattening technique in open estuarine area which was taken on lease and practiced it in depth of not more than 1.5 meter. One side of bamboo pole was tied to the last legs of crab with help of rope and the opposite side was attached to a float, which is a square shaped thermocole, for easy detection. With the help of small boat, every crab can be checked and released again in the estuary and due to natural environment crabs look healthy and colourful. Intensive feeding promotes growth to realize early returns as fattening occurs within 15-20 days.

Practiced by: Anil Anthony, Athavada bajar, Ratnagiri-415612

Documented by: Shrikrishna P. Hotekar, M.F.Sc. 2011-12 College of Fisheries,Ratnagiri, Email: hotekarsp@gmail.com

Mrs. Apurva N. Swant, 1 st year Phd Aquaculture, College of Fisheries, Ratnagiri

Email: archana114@rediffmail.com

Dr. Suresh. D. Naik, Associate professor, HOD, FPTM, College of Fisheries, Ratnagiri

Email: naiksdn2003@yahoo.co.in

Towards the end it can be said that grassroot innovations refer to innovative solutions that are developed and implemented at the local level, often by communities or individuals, rather than by large organizations or governments. These innovations are typically driven by the need to address a particular problem or challenge and are often low-cost, simple, and sustainable.

Grassroot innovations can encompass a wide range of solutions, including new technologies and community initiatives. They often arise in response to a lack of existing solutions, and they are characterized by their ability to quickly adapt to changing conditions and to be scaled up or down as needed.

Grassroot innovations are an important source of new and innovative solutions to the world's most pressing problems, and they often provide an effective alternative to top-down approaches.

GRASSROOT INNOVATIONS ARE IMPORTANT FOR SEVERAL REASONS:

They Address Local Needs:

Grassroot innovations are often developed in response to specific problems faced by a particular community or group of people. By addressing these local needs, these innovations can provide effective solutions that are tailored to the particular context.

They Are Cost-Effective:

Many grassroot innovations are simple and low-cost, making them accessible to communities and individuals who might not otherwise have access to solutions. This can lead to increased access to resources and services, especially in rural or underserved areas.

They Promote Empowerment:

By developing and implementing their own solutions, communities and individuals are able to take control of their own development and to address their own challenges. This can lead to increased empowerment and self-sufficiency, as well as a sense of pride and ownership in the solutions they have created.

They Drive Innovation:

Grassroot innovations can lead to new and innovative solutions, as communities and individuals are able to draw on their own experiences, knowledge, and creativity to address problems in new and creative ways. This can help to drive innovation and progress, both at the local and global level.

They are Scalable:

Grassroot innovations can be scaled up or down as needed, making them well- suited to addressing problems and challenges in a wide range of contexts. This can help to ensure that solutions are sustainable and can be adapted to changing conditions over time.

Grassroot innovations are an important source of innovation and progress, and they play a critical role in addressing some of the world's most pressing challenges and in driving sustainable development.

Grassroot innovations can play a significant role in the fisheries and aquaculture sector, by addressing specific challenges and promoting sustainable development.

Some examples of grassroot innovations in fisheries and aquaculture sector have been presented in this resource book as they have a potential to play a crucial role in addressing the challenges faced by small-scale fishers and communities.

These innovations can improve the efficiency and sustainability of fishing operations, increase the economic and social benefits of fishing, and contribute to the conservation of marine ecosyste Some of the key benefits of grassroot innovations in the fisheries sector are:

Improved Livelihoods:

Grassroot innovations can provide small-scale fishers with new opportunities to increase their income and improve their standard of living. For example, the development of new fishing gear and techniques can help fishers to catch more fish, sell their catch for a higher price, and reduce their costs.

Increased Efficiency:

Innovations at the grassroot level can help to reduce the effort and costs associated with fishing, making it more economically viable for small-scale fishers. For example, the development of low-cost fishing gear can reduce the cost of fishing, while new techniques and technologies can help fishers to locate and target fish more effectively.

Enhanced Sustainability:

Grassroot innovations can help to promote the sustainable use of marine resources, by reducing the impact of fishing on the environment and conserving fish stocks. For example, the development of selective fishing gear can help to reduce the bycatch of non-target species, while the use of eco-friendly fishing techniques can minimize the damage to habitats.

Empowerment of Communities:

Grassroot innovations can help to empower small-scale fishers and fishing communities by giving them greater control over their livelihoods and resources. By providing access to new technologies, techniques, and information, grassroot innovations can help communities to take charge of their own development and make decisions that are in their best interests.

Grassroot innovations in the fisheries and aquaculture sector have the potential of having significant impact on the lives of small-scale fishers and communities, by improving their livelihoods, increasing efficiency, promoting sustainability, and empowering communities.

India has been the leader in developing grass root innovations. Like Mahatma Gandhi ji, the leader of India's independence movement, believed in the power of local initiatives and grassroots innovations to bring about social and economic change. He encouraged people to be self-sufficient and to work together in their communities to solve local problems and improve their lives.

Gandhi ji believed that the solutions to many of India's problems could be found at the grassroots level, and that the government should not try to impose solutions from above. Instead, he emphasized the importance of empowering individuals and communities to take action and solve problems on their own.

In this context, Gandhi ji saw grassroots innovations as a way for people to use their own skills, resources, and creativity to address local problems and improve their communities. He encouraged people to be proactive and to work together to find practical solutions to the challenges they faced.

Here is one quote from Gandhi ji that highlights his views on grassroots innovations: "Strength does not come from physical capacity. It comes from an indomitable will." In this quote, Gandhi ji emphasizes the power of individual and collective willpower to bring about change, even in the face of difficult challenges. This idea is closely related to his views on grassroots innovations, as it emphasizes the importance of local initiative and collective action in driving social and economic progress.

Dr. Anil Gupta who is an Indian academic and social entrepreneur who is the founder of the National Innovation Foundation (NIF), a government-supported organization that works to promote and support grassroots innovations in Ahemdabad has been a strong advocate for the importance of grassroots innovations in driving social and economic change, and has dedicated much of his career to identifying and supporting innovative solutions to problems at the local level.

According to him, grassroots innovations are crucial for addressing the needs and challenges of rural and remote communities, which are often overlooked by mainstream development progra He argues that these innovations have the potential to create new opportunities for entrepreneurship and economic growth, and to improve the quality of life for people in rural areas.

Here is one quote that highlights his views on grassroots innovations:

"Grassroot innovations are locally grounded solutions to locally relevant proble They are innovations that have been created from the bottom up, often by individuals or groups working outside of mainstream systems and institutions. These innovations can be small and incremental, but they can also be transformative and have a big impact."

In this quote, it has been emphasized about the importance of locally grounded solutions in addressing local problems, and the potential of grassroots innovations to bring about significant change. He also highlights the fact that these innovations often come from individuals or groups working outside of mainstream systems, emphasizing the importance of supporting and nurturing such initiatives in order to create a more inclusive and sustainable development model.

In India, the Government has taken a number of initiatives like providing support, recognition, and dissemination opportunities for innovative technologies and practices developed by individuals and communities, to promote and support grassroots innovations, recognizing the important role that the innovations can play in driving social and economic change.

Indian Council of Agricultural Research (ICAR) has played an important role in promoting and supporting grassroots innovations in the agricultural sector in India.

Grassroot innovations in agriculture are crucial for addressing the challenges faced by small-scale farmers and rural communities, and ICAR recognizes their potential to improve the productivity, profitability, and sustainability of agriculture in India.

To support grassroots innovations in agriculture, ICAR has established a number of programmes and initiatives that promote the development and dissemination of innovative technologies, practices, and techniques.

For example, ICAR has established a network of Krishi Vigyan Kendras (KVKs) that work at the local level to transfer agricultural technology to farmers, and to promote innovation and entrepreneurship.

ICAR also works to identify and support innovative ideas and technologies that have been developed by farmers and rural communities, and to provide these innovations with the support and recognition they need to reach a wider audience.

ICAR plays a key role in promoting and supporting grassroots innovations in the agricultural sector in India. By providing support, recognition, and dissemination opportunities for innovative technologies and practices developed by farmers and rural communities, ICAR is helping to create a more inclusive and sustainable model of agricultural development in India.

The ICAR-Central Institute of Fisheries Education (CIFE), Mumbai has worked in this area as grassroot innovations in the fisheries sector are crucial for addressing the challenges faced by small-scale fishers and fishing communities, and CIFE recognizes their potential to improve the productivity, profitability, and sustainability of the sector.

To support grassroots innovations ICAR-CIFE works with local fishing communities to develop and promote alternative livelihood options. CIFE has also taken the initiative to identify and support innovative ideas and technologies that have been developed by fishers and fishing communities, and to provide these innovations with the recognition they need to reach a wider audience.

I think frugality drives innovation, just like other constraints do. One of the only ways to get out of a tight box is to invent your way out.

Jeff Bezos

Our future growth relies on competitiveness and innovation, skills and productivity... and these in turn rely on the education of our people.

Julia Gillard

The five essential entrepreneurial skills for success: Concentration, Discrimination, Organization, Innovation and Communication.

Harold S. Geneen

Innovation comes from the producer - not from the customer.

W. Edwards Deming

Innovation distinguishes between a leader and a follower.

Steve Jobs

Without change there is no innovation, creativity, or incentive for improvement. Those who initiate change will have a better opportunity to manage the change that is inevitable.

William Pollard

Innovation is the specific instrument of entrepreneurship. The act that endows resources with a new capacity to create wealth.

Peter Drucker

The Indian Government's Initiatives for Grassroot Innovations

In the context of India's diverse socio-economic landscape, grassroots innovations have emerged as a vital component for sustainable development. These innovations, typically developed by local communities, address specific needs and challenges faced at the local level. Recognizing the importance of grassroots innovations, the Indian government has initiated various policies, programmes, and institutions to increase creativity, entrepreneurship and technological advancement at the grassroots level.

One of the primary ways the Indian government supports grassroots innovations is through the establishment of dedicated institutions and funds aimed at promoting research and development. The National Innovation Foundation (NIF), set up by the Government of India, plays a pivotal role in this regard. NIF is tasked with scouting, sustaining and scaling up grassroots innovations. It aims to create a robust ecosystem that encourages innovators, especially in rural areas, to bring their ideas to fruition. Through various initiatives, NIF offers financial support, technical assistance and market linkages to grassroots innovators, helping them transform their concepts into viable products.Moreover, the NIF collaborates with various educational and research institutions to facilitate innovation at the grassroots level. By organizing innovation contests and exhibitions, the NIF provides a platform for local innovators to showcase their work, gain recognition, and connect with potential investors and collaborators. These efforts not only highlight the creativity and ingenuity of grassroots innovators but also inspire others in the community to engage in innovative activities.

Another significant initiative by the Indian government is the establishment of the Atal Innovation Mission (AIM), which aims to promote a culture of innovation in the country. AIM focuses on fostering innovation and entrepreneurship across various sectors, including agriculture, health, education, and more. It provides financial support to startups and grassroots innovators through its various programmes. The mission encourages collaborations between educational institutions, research organizations, and local communities to develop and implement innovative solutions to pressing challenges.

The Government of India has also launched the Startup India initiative, which is designed to promote entrepreneurship across the country. This initiative includes various measures to support grassroots innovations, such as simplifying regulatory processes, providing funding opportunities, and offering mentorship programmes. The government has set up numerous incubators and accelerators, particularly in rural and semi-urban areas, to nurture innovative ideas and transform them into successful ventures. By doing so, the government is not only promoting grassroots innovations but also contributing to job creation and economic development in underserved regions.

In addition to these initiatives, the Indian government recognizes the importance of intellectual property rights (IPR) in encouraging innovation. To this end, the government has implemented various schemes to help grassroots innovators protect their ideas and inventions.

The role of local self-governments in promoting grassroots innovations cannot be overlooked. Panchayati Raj institutions, which serve as the grassroots-level governance framework in rural India, are increasingly becoming active participants in fostering innovation. The government encourages these local bodies to support innovative projects by providing them with funding and resources.

Moreover, the Indian government has recognized the significance of traditional knowledge and practices in fostering grassroots innovations. Programmes aimed at documenting and preserving traditional knowledge systems have been initiated to promote sustainable practices that have been developed over centuries. This not only helps in conserving biodiversity but also empowers local communities by validating their knowledge and practices. By integrating traditional knowledge with modern scientific approaches, the government aims to create a holistic framework for innovation that respects cultural heritage while addressing contemporary challenges.

The government also promotes skill development as a means to enhance grassroots innovations. Various skill development programmes are aimed at equipping individuals with the necessary skills to innovate and create. The Skill India Mission focuses on providing training in various sectors, enabling individuals, especially youth, to harness their creativity and contribute to local development. By investing in skill development, the government is not only fostering innovation but also enhancing the employability of the local workforce.

Publicprivate partnerships (PPPs) have emerged as a crucial strategy for promoting grassroots innovations. The Indian government actively collaborates with private sector players, NGOs, and academic institutions to create a conducive environment for innovation. These partnerships facilitate knowledge sharing, resource mobilization and capacity building, ultimately benefiting grassroots innovators. By leveraging the strengths of both the public and private sectors, the government aims to create a more sustainable and inclusive innovation ecosystem.

The significance of grassroots innovations extends beyond economic development; they also play a critical role in addressing social issues. The Indian government has initiated programmes that focus on social innovations aimed at improving the quality of life for marginalized communities. These initiatives often involve local innovators who design solutions tailored to the unique challenges faced by these communities. By addressing issues such as sanitation, healthcare, and education through grassroots innovations, the government seeks to uplift the most vulnerable sections of society.

In addition to these measures, the Indian government emphasizes the importance of awareness and capacity building among local communities. Various outreach programmes are conducted to educate communities about the importance of innovation and entrepreneurship. By creating awareness about available resources, funding opportunities, and support mechanisms, the government empowers local innovators to take charge of their development. Workshops, seminars, and training sessions are organized to equip individuals with the knowledge and skills needed to transform their ideas into viable solutions.

Furthermore, the government actively encourages participation from women and marginalized groups in grassroots innovations. Special initiatives are designed to promote women's entrepreneurship and innovation, recognizing their potential to drive economic growth and social change. By providing targeted support and resources, the government aims to create an inclusive environment that encourages diverse perspectives and contributions to the innovation landscape.

The role of education in promoting grassroots innovations is also significant. The government has recognized the need to integrate innovation into the educational curriculum at all levels. By fostering a culture of creativity and critical thinking among students, the government aims to nurture the next generation of innovators. Initiatives like the National Educational Policy (NEP) emphasize the importance of experiential learning and interdisciplinary approaches, which can significantly enhance the innovation potential of students.

Research and development (R&D) activities are vital for sustaining grassroots innovations. The Indian government has increased its investment in R&D to support innovative projects that address local challenges. By collaborating with research institutions and universities, the government facilitates the transfer of knowledge and technology to grassroots innovators. This partnership not only strengthens the innovation ecosystem but also ensures that the solutions developed are scientifically sound and sustainable.

Additionally, sharing success stories and best practices can inspire other innovators and create a culture of continuous improvement which is one of the objectives of this book. The Indian government has undertaken a multifaceted approach to promote grassroots innovations, recognizing their significance in driving sustainable development and social empowerment. Through various initiatives, institutions and policies, the government aims to create an enabling environment for local innovators to thrive. By fostering creativity, entrepreneurship and collaboration, the government seeks to harness the potential of grassroots innovations to address the pressing challenges faced by the nation. This comprehensive support not only empowers local communities but also contributes to the overall growth and prosperity of the country. As India continues to evolve, grassroots innovations will play an increasingly important role in shaping its future, making it imperative for the government to sustain and enhance its efforts in this critical area.

Role of Indian Council of Agricultural Research in Grassroots Innovations

The Indian Council of Agricultural Research (ICAR) has been a cornerstone of agricultural research and development in India since its establishment. Its mission extends beyond scientific research; it aims to foster sustainable agricultural practices, enhance food security, and improve the livelihoods of farmers. ICAR has played a role is its engagement with grassroots innovations. Grassroots innovations are locally developed solutions to agricultural challenges, often created by farmers themselves.

Grassroots innovations refer to the creative solutions that emerge from the experiences of local communities. In agriculture, these innovations can range from improved seed varieties to novel farming techniques and resource management strategies. Unlike topdown approaches, grassroots innovations are often characterized by their adaptability, cost effectiveness and sustainability. They arise in response to specific challenges faced by farmers, reflecting local knowledge and cultural practices.

ICAR recognizes that for grassroots innovations to thrive, farmers must be equipped with the necessary skills and knowledge. Therefore, capacity building is a critical component of ICAR's strategy. Through workshops, training programmes and demonstrations, ICAR helps farmers understand

and adopt new techniques and practices. This training often includes topics such as sustainable farming practices, integrated pest management, and soil health management.

ICAR has made efforts to document and support indigenous technologies that have evolved over generations. By recognizing the value of local knowledge systems, ICAR encourages the use of traditional practices alongside modern scientific research. This approach not only preserves cultural heritage but also enhances the resilience of farming systems. For example, the development of drought-resistant crop varieties often involves incorporating traditional knowledge about local seed selection and cultivation practices.

ICAR plays a crucial role in advocating for policies that support grassroots innovations. One of the essential roles is to document and disseminate successful grassroots innovations. By creating databases and publications that showcase these innovations, it ensures that valuable knowledge is shared across regions and communities. This documentation also helps in creating awareness about successful practices that can be replicated elsewhere. Moreover, the dissemination of this information through various channels, including workshops, seminars, and digital platforms, amplifies the reach and impact of grassroots innovations. To understand the effectiveness of grassroots innovations, it emphasizes the need for rigorous evaluation. This includes assessing the social, economic, and environmental impacts of these innovations on farming communities.

Role of Women in Grassroots Innovations

Women play a crucial role in Indian agriculture and fisheries, significantly contributing to food production, resource management, and community welfare. Their involvement is particularly evident in grassroots innovations, where local knowledge and practices often lead to sustainable solutions for various challenges. In agriculture, women constitute nearly 33% of the agricultural labour force, engaging in activities such as sowing, weeding, harvesting, and post-harvest management. Their handson experience allows them to identify problems and develop innovative solutions tailored to their specific contexts.Women possess traditional knowledge about local crops, medicinal plants, and sustainable farming practices, local fishes, sustainable fishing practices which is often passed down through generations. This knowledge includes insights into crop rotation, pest management, and soil health, making their contributions essential for maintaining agricultural diversity. Women also play a vital role in seed selection and preservation, often managing diverse seed banks within their communities.Furthermore, many womenled initiatives focus on sustainable practices, such as organic farming and agro-ecology. For instance, women in various regions have developed organic pest control methods using locally available materials. These innovations not only enhance productivity but also promote environmental sustainability.

In the fisheries sector, women are extensively involved in pre harvest, harvest and post harvest activities. They often manage small scale fish processing units, employing traditional methods to enhance the quality and shelf life of fish products. This role not only increases household income but also improves food security for their families. Women also contribute to community management of fisheries resources, participating in decision-making processes related to resource use, conservation practices, and conflict resolution. Their insights into local ecosystems are crucial for developing sustainable fishery practices.Moreover, women have pioneered various aquaculture innovations, such as integrated farming systems that combine fish farming with agriculture. These practices optimize resource use and increase productivity, showcasing women's ability to adapt to changing environmental conditions.

Role of women in grassroots innovations in Indian agriculture and fisheries is indispensable. Their unique perspectives, traditional knowledge and innovative practices significantly contribute to sustainable development.

Notes